WEYERHAEUSER ENVIRONMENTAL BOOKS

Paul S. Sutter, Editor

WEYERHAEUSER ENVIRONMENTAL BOOKS explore human relationships with natural environments in all their variety and complexity. They seek to cast new light on the ways that natural systems affect human communities, the ways that people affect the environments of which they are a part, and the ways that different cultural conceptions of nature profoundly shape our sense of the world around us. A complete list of the books in the series appears at the end of this book.

CHARGED

*A History of Batteries and Lessons
for a Clean Energy Future*

JAMES MORTON TURNER

Foreword By Paul S. Sutter

UNIVERSITY OF WASHINGTON PRESS
Seattle

Charged: A History of Batteries and Lessons for a Clean Energy Future is published with the assistance of a grant from the Weyerhaeuser Environmental Books Endowment, established by the Weyerhaeuser Company Foundation, members of the Weyerhaeuser family, and Janet and Jack Creighton.

Composed in Minion Pro, typeface designed by Robert Slimbach

UNIVERSITY OF WASHINGTON PRESS
uwapress.uw.edu

THE LIBRARY OF CONGRESS HAS CATALOGED THE HARDCOVER EDITION OF THIS BOOK AS FOLLOWS:
Names: Turner, James Morton, 1973– author.
Title: Charged : a history of batteries and lessons for a clean energy future / James Morton Turner.
Description: Seattle : University of Washington Press, [2022] | Series: Weyerhaeuser environmental books | Includes bibliographical references and index. | Identifiers: LCCN 2021037424 (print) | LCCN 2021037425 (ebook) | ISBN 9780295750248 (hardcover) | ISBN 9780295750262 (ebook)
Subjects: LCSH Electric batteries.
Classification: LCC TK2896 .T87 2022 (print) | LCC TK2896 (ebook) | DDC 621.31/242—dc23/eng/20211110
LC record available at https://lccn.loc.gov/2021037424
LC ebook record available at https://lccn.loc.gov/2021037425

Paperback ISBN 9780295752181

♾ This paper meets the requirements of ANSI/NISO z39.48-1992 (Permanence of Paper).

For Cole, Liam, and August

CONTENTS

FOREWORD

What's the Matter with Batteries?

PAUL S. SUTTER

When I was in graduate school in the early 1990s, there was a crystalline piece of graffiti stenciled on the sidewalk in front of the campus library. It read, simply, "Ecology tip: consume less." This subversively friendly counsel paired the banal genre of eco-advice with a radical message: people (and particularly wealthy people) used far too many resources, and our consumer habits were driving ecological destruction. There was nothing novel about that advice; the call to reduce the ecological footprint of the developed world had anchored global environmental politics for a couple of decades by that point. But I appreciated the economy of the message, and it became my environmental mantra. What I appreciated less well, at first, was the sardonic quality of the message, the mocking sense that its easy guidance was going to be exceptionally difficult to execute. The solution to global ecological destruction, it intimated, would not involve the stuff of "tips": a lifestyle tweak or faithful adherence to green consumerism. Consuming less was going to hurt, both as an individual practice and as a structural fix. Over the last quarter century, as we have waded deeper into a planetary crisis, this apparent solution to our environmental problem—consuming less—has remained as clear as our incapacity to achieve it. The brilliance of this small piece of public art, I came to realize, was as much in its diagnosis of the wicked problem we face as in its purported solution.

More than anything I have read, James Morton Turner's compelling new book, *Charged: A History of Batteries and Lessons for a Clean Energy Future*, has shaken my environmentalist faith that consuming less is a feasible path forward, particularly when it comes to countering climate change. The world we live in requires energy, and while we should certainly pursue conservation and other efficiencies, we cannot assume that decarbonization will be achieved through reduced energy consumption. Instead, the urgency of the climate crisis demands that we rapidly pivot to a clean energy economy to meet our present and future energy needs. Turner is an advocate for this transition, but his vital caution is that our clean energy future will have dramatic material consequences. It will require new technologies and infrastructures: electric vehicles by the hundreds of millions, vast solar arrays and fields of wind turbines, a new grid to deliver this sustainable energy, and, critically, the scaling up of the production of batteries to store this energy and mobilize its power. This rapid transition is possible, but it will involve the consumption of natural resources on a massive scale, and it will have consequences for human justice as well as environmental sustainability.

To assess the material consequences of our aspirational clean energy future, Turner focuses his analysis on the history of the batteries that powered the twentieth century. Batteries are such a mundane technology—whether they be the lead-acid batteries in conventional automobiles, the AA batteries in flashlights, or the lithium-ion batteries in smartphones—that we tend not to think about them. They are, he points out, a persistent black box that we take for granted without understanding their inner workings. One of the great virtues of *Charged* is the education you will receive in how batteries work and how they have evolved over time. Battery literacy, or what Turner more broadly calls "industrial ecological literacy," is so important to him that at several points he recounts building his own rudimentary batteries or, in other cases, taking batteries apart. These exercises show the reader what batteries are and how they work, but they are also metaphors for what Turner is trying to accomplish more broadly: to open up the black box and consider how the sourcing and production of batteries connect us to many other communities and landscapes.

Several lessons flow from cracking batteries open to get a better look. While batteries themselves are not primary sources of energy, they have been critical to storing and mobilizing energy produced by other sources, and they will be critical to our use of wind, solar, hydro, and other renewable energy sources, none of which are as portable or reliable as fossil fuels. As purveyors of energy, Turner tells us, batteries are inefficient, but we value them for the quality of energy that they deliver. Batteries provide energy that is cordless, reliable, and portable; they are the unsung heroes of what Turner calls our "culture of mobility." To maintain this modern culture of convenience while weaning ourselves from fossil fuels, we are going to need *a lot* of batteries.

A second lesson is that batteries are materials-intensive technologies with their own suite of environmental costs. But because we pay so little attention to batteries, and particularly where they come from, this can be hard to see. Depending on what type of batteries we are talking about, they might require (or have required in the past) lead, sulfuric acid, mercury, manganese, zinc, steel, carbon, graphite, ammonium chloride, potassium hydroxide, cadmium, lithium, nickel, or cobalt. Batteries are of the earth; they are the stuff of mines and metallurgy and energy-intensive materials processing, and when their useful lives come to an end, they must either be recycled or returned to the earth. Thus, while batteries will be critical to liberating us from our fossil fuel dependence, we will be trading one set of materials dependencies for another as we rapidly scale up their production. Leaving fossil fuels in the ground, in other words, is going to involve pulling a lot of other resources out of it.

A third critical lesson of Turner's materialist approach is that the industrial life cycles of batteries have significant human costs, which are unevenly spread across the globe. The materials used to produce batteries have to be mined or recycled, and someone must do that dirty and often dangerous work. These materials then must be processed and assembled, with the workers engaged in these tasks facing potentially harmful exposures. The minerals necessary for battery production, which are increasingly diverse, are not uniformly available but favor certain geographical locations over others, meaning that access to them has been and will continue to be uneven and geopolitically complex. Moreover, the optimal conditions for inexpensive mining, processing, and recycling often favor nations with weak labor laws and environmental regulations. As we necessarily scale up the production of batteries to meet future energy storage and mobility needs, Turner insists that we keep our eyes on these differential geographies and their injustices. Understanding the full scope of the industrial lifecycle of batteries will be critical to achieving policies that promote a just as well as a clean energy transition.

As should be clear by now, Turner is not an ecomodernist or technological optimist; he does not truck in Promethean narratives in which batteries solve all of our energy problems. Nor is he a skeptical naysayer who finds joy in muddying the utopian visions of clean energy advocates and Green New Dealers. He is, above all, a clear-eyed guide to the energy transition that we need to make, a careful historical thinker wary of simple solutions or technological panaceas, a cautious advocate for our battery-powered future who insists we learn from our battery-powered past.

In providing such guidance, Turner also takes aim at how traditional environmental politics will need to adjust to this new energy reality. For a readership with environmentalist sympathies, this may be the most challenging and important part of *Charged*. Turner insists that we cannot simultaneously embrace clean energy technologies and oppose the mines and manufacturing facilities that will make them possible; we cannot scale up renewable energy generation, and the production of batteries that that will require, if we still cling to soft energy paths and "small is beautiful" ideals; and we cannot rapidly shift to green energy sources and technologies if we continue to manifest deep suspicions about synthetic materials or assume that we can recycle our way out of the material demands required by this transition. Our continued hope for a return to the simple life—a future in which the purported solution to our environmental problems simply involves consuming less—has gotten in the way of our ability to think clearly about what Turner calls "our deepening and inescapable 'entanglement' in the material world and the cascade of consequences that has for both human and environmental health." Instead, Turner advances a "material environmentalism" that calls for a new sort of stewardship of natural resources and human communities. In doing so, *Charged* offers a profound and unflinching examination of what sustainability will look like across the twenty-first century.

Ecology tip: read this book.

ACKNOWLEDGMENTS

I began researching this book in 2010. Many times along the way, I nearly gave up on this project: I was overwhelmed by the opaqueness of the modern minerals industries, the complexities of battery chemistry, and the injustices of resource extraction both past and present. But as the clean energy transition began to gain momentum, and the role of batteries in enabling it became ever more apparent, this project became increasingly urgent. After more than a decade of research, I wrote most of this book between April 2020 and February 2021.

During this book's decade-long windup, I received an extraordinary amount of help from people in the battery and materials industries, colleagues, and students. Many people in industry provided interviews and tours of smelters, recyclers, and other facilities. I met with some of them so long ago that they've likely forgotten my visit. But those conversations and experiences played a key role in shaping this project. Thank you to Ashish Bhandari, Khush Marolia, Alain Vassart, Bob Finn, Carl Smith, Dick Amistadi, Hans Craen, Dan Kinsbursky, Marc Boolish, Paul Johnson, Jean-Yves Huot, Daniel Leach, Robert Scarr, Ray Balfour, Todd Coy, Terry Telzrow, and Art Hebrank, among others.

I also learned a tremendous amount from those in government, academia, and the nonprofit sector who have been concerned with the trade in materials and management of batteries at end of life. My thanks to those who took the time to speak with me, including Laura Coughlan, Rob D'Arcy, Perry Gottesfeld, Garth Hickle, Ruska Kelevska, Elsa Olivetti, Heidi Sanborn, Robin Schneider, Linda Taylor, Jean Wagenius, and Tim Whitehouse.

This book considers how the history of batteries intersects with places around the world, including Siberia, China, the Democratic Republic of the Congo, Peru, Australia, Chile, Argentina, and Missouri, among others. I visited few of those places in person. I am sure the book would have been richer had my travels been more extensive. I know this because of the tremendous amount I did learn from visits to communities in Missouri and East Los Angeles, California, which are on the front lines of the lead industry. My thanks to the people in those communities who shared their stories with me, including Dale Brooks, Brenda Browning, Maya Golden-Krasner, Milton Nimatuj, Gerty and Larry O'Leary, Idalmis Vaquero, and Leslie and Jack Warden.

My students inspired this book, and many of them played important roles in bringing it to fruition. My thanks to Wenyin Cao, Kelsey Dunn, Hannah Flesch, Yasmin Kaiser, Wangũi wa Kamonji, Mackenzie Klema, Aynsley Kretschmar, Maddie Lee, Sarah McBride, Marissa Menzel, Leah Nugent, Eva Paradiso, Eve Silfanus, Anastasia Thayer, and Madhur Wale for your help researching and fact checking this book, and your advice on its content.

This is a project that has stretched my training as an environmental historian. My colleagues at Wellesley College have been generous and patient with their advice and expertise in fields ranging from chemistry to political science. My thanks to Rebecca Belisle, Dan Brabander, Beth DeSombre, Nolan Flynn, Alden Griffith, Tom Hodge, Lidwien Kapteijns, Erich Matthes, and Rachel Stanley. I've also benefited greatly from the help of many fellow historians, including Megan Black, Connie Chiang, Leif Fredrickson, Andrew Isenberg, Adam Rome, Chris Sellers, Teresa Spezio, Jennifer Thomson, and Chris Wells, among others. Several individuals with expertise in batteries and the devices they power were extraordinarily generous in reviewing portions of the manuscript for its technical content. My thanks to Linda Gaines, David Morrow, Elsa Olivetti, and Jay Roudebush. Despite this assistance, I remain responsible for the errors (and I'm sure there are some!) that remain.

My thanks to the entire team at the University of Washington Press that shepherded this book to publication. I greatly appreciate the outsider readers who read more than one version of this book. I'm especially grateful for Paul Sutter's patience, enthusiasm, and critical insights for the past decade, as I've figured out how to write this book. It wouldn't have happened without his advice and encouragement. I'm deeply grateful to Andrew Berzanskis, who helped shape the final manuscript and provided support at just the right moments to get this project over the finish line. And Susan Murray's wonderful copy editing improved the book's clarity and prose.

This work has been made possible by grant support from the National Science Foundation (Social and Economic Sciences Award 1230521) and the National Endowment for the Humanities (FB-56565–12). It has also benefited from the support of the Newhouse Center for the Humanities, the Knapp Social Science Center, the Frost Center for the Environment, and the Environmental Studies Department at Wellesley College.

Finally, I offer my deep thanks to my family for putting up with my seemingly endless fascination with batteries.

CHARGED

INTRODUCTION

BATTERIES INCLUDED

WHEN I DRIVE WEST ACROSS THE GREAT PLAINS, I USUALLY KEEP MY eyes glued to the horizon. I am looking for the first signs of the Rocky Mountains rising in the distance. The last time I made that drive, several summers ago, the wind turbines caught me by surprise. They made for a striking vista. As I sped along Interstate 70, I passed hundreds of wind turbines, some of which loomed high above the interstate and others that spun far in the distance. I had not visited the region for more than a decade, and during that time Colorado had positioned itself on the leading edge of a clean energy transition. In 2020, wind power supplied nearly one-fourth of Colorado's electricity needs.[1]

Most of the electricity the wind turbines generated likely traveled the same way I was—west, toward Denver, Boulder, and other cities along the Front Range. And, as I approached the Front Range, I saw trains loaded with turbine blades heading east, destined for new wind farms going up on the plains. More than seven thousand people worked in the Colorado wind industry, many of them at one of the Colorado factories that manufactured towers, blades, or nacelles (which house gears and generators atop the towers).[2] For proponents of a clean energy future, this is what the twenty-first century is meant to look like.

The turn toward renewable energy gained surprising momentum at the start of the twentieth-first century. In 2000, government analysts projected that renewable electricity generation would grow at less than 0.5 percent per year to reach 400 billion kilowatt-hours by 2020.[3] They were wrong. Renewable electricity generation hit that target in 2010 and grew at a rate of more than 10 percent per year after 2010 to reach more than 800 billion kilowatt-hours in 2020.[4] Renewable electricity generation exceeded coal generation in 2020—a turnabout that energy forecasters did not anticipate in 2000.[5] Even the *Wall Street Journal* marveled at the pace of change: "What seemed like an impossibility just a decade ago—the displacement of fossil fuels from the U.S. power system, if not the world's—is increasingly a reality."[6]

The signs of a growing clean energy revolution can be seen elsewhere in the United States too. From California to Texas to New York, schools, office parks, municipal buildings, and homes are being designed or retrofitted as "zero energy buildings." By pairing on-site solar arrays with high-efficiency heating and cooling systems, effective insulation, and other energy-saving measures, a net-zero building can generate as much energy as it uses over the course of a year. Starting in 2020, net-zero building design became standard practice in California. The state mandated that all residential construction, including all new homes and some renovations, meet a state-specific net-zero energy standard.

Another sign of change was tucked into those building codes: standards for making new construction "EV-ready" in anticipation of a future when most homes will have an electric car charging in the driveway or garage. In the 2010s, Tesla became synonymous with electric vehicles, gaining a toehold with premium vehicles, such as the Model S and Model 3. But at the start of the 2020s, conventional automakers began to bet big on an electric future too. In 2021, Ford announced an electric version of the F-150, the world's most popular pickup truck. Some of the world's largest automakers, including General Motors and Volkswagen, committed to phasing out gas-powered cars. Volkswagen promised to do so by 2026. And governments around the world set goals demanding that automakers make good on those promises. The United Kingdom, Austria, and China were among the first to adopt timelines for phasing out conventional automobile sales. Norway's ban takes effect in 2025.[7] The United States is aiming to make 50 percent of new car sales electric vehicles by 2030.

For technological optimists, these advances signaled the beginnings of a clean energy revolution that many believe will transform the twenty-first century, driven by a large-scale shift away from oil, gas, and other fossil fuels and toward an electrified future, where cars, buildings, and industry will all be powered by renewable electricity. It is a transition that has been a long time coming.

In the 1970s, when the energy crisis gripped the United States and oil prices soared, Amory Lovins described such an alternative energy future in *Soft Energy Paths*. Lovins, an early proponent of energy efficiency and distributed energy generation, argued that if the political will could be mustered, a "largely or wholly solar economy can be constructed in the United States with straightforward soft technologies that are now demonstrated and now economic or nearly economic."[8]

Such optimism faded during the 1980s, when the energy crisis eased and renewable energy prices remained high. But it was rekindled at the start of the twenty-first century, as concerns about global warming mounted. In 2004, Stephen Pacala and Robert Socolow outlined a strategy for tackling climate change that did not require any "revolutionary technology." The two Princeton-based scientists explained that "humanity can solve the carbon and climate problem in the first half of this century simply by scaling up what we already know how to do."[9] In the 2010s, Stanford University engineer Mark Jacobson and his team modeled how cities, states, and countries could transition entirely to renewable energy sources by 2050 by pairing large-scale renewable energy generation from wind, solar, and hydroelectricity with improvements in electricity distribution, energy-storage technologies, and other efficiency improvements.[10] They projected Colorado could transition to 100 percent renewable energy by 2050, meeting 55 percent of its energy needs with wind power, 37 percent with solar power, and the balance from hydroelectricity and geothermal plants.[11]

Important to such visions of a clean energy future is more than just a set of technologies, however. They are also predicated upon a changing ethos for modern sustainability advocacy. In place of an earlier generation of environmental activism, which emphasized limits, compromise, and sacrifice, a new generation of green visionaries sees in technological innovation the possibilities for a more sustainable future predicated on innovation, plenty, and growth.[12] Tesla exemplifies this vision. Its mercurial chief executive, the serial entrepreneur Elon Musk, has spun a sustainability narrative that celebrates high performance and design. The Tesla vision is manifest in glossy renderings of the future suburban home, shingled in elegant solar roof tiles, sporting sleek battery packs for energy storage, and housing a Tesla vehicle in the garage. It gives no heed to potential limits on technology, consumerism, or population—concerns that preoccupied environmentalists in the

1970s. Instead, its ultimate goal is "infinitely scalable clean energy generation and storage products."[13]

Other sustainability thinkers echoed such aspirations. In 2006, Alex Steffen, an emerging leader of a new generation of green visionaries, acknowledged that "it may seem impossibly far away, but on days when the smog blows off, you can already see it: a society built on radically green design, sustainable energy, and closed-loop cities; a civilization afloat on a cloud of efficient, non-toxic, recyclable technology." In his words, "That's a future we can live with."[14] By 2015, a group of leading environmental scholars and activists joined together, branding themselves as "ecomodernists" who shared a "vision for putting humankind's extraordinary powers in the service of creating a good Anthropocene." In place of environmentalism's longstanding discomfort with technology, ecomodernists saw in technology and design the best chance to meet the environmental crisis ahead while protecting a thriving environment and human society.[15]

As the world begins to chart its way toward this clean energy future, this book takes a step back. It explores the relationship between environmentalism, sustainability, and technology at a moment when such technological optimism is ascendant. To do so, it focuses on one technology that is vitally important to unlocking a clean energy future: the battery. Batteries will help store electricity from solar panels and wind turbines. Batteries will help improve the reliability, versatility, and efficiency of the electrical grid. And batteries will power a new generation of zero-emissions vehicles, from cars to bicycles to airplanes. As battery production continues to scale up, industry analysts are forecasting that battery prices will continue to fall—dropping below $100 per kilowatt-hour in 2024. At that price, electric cars should be cheaper than gasoline cars and utility-scale electricity storage—to meet periods of peak demand and even out the ups and downs of renewable electricity generation—increasingly viable. To hasten this transition, companies like Tesla are building battery "Gigafactories" to meet the growing demand.

But as skeptics of a technology-driven clean energy transition like to remind environmentalists, batteries have a dirty little secret. Although batteries may make it possible to scale up renewable-energy sources, deploy electric cars, and wean the world off fossil fuels, helping to solve global warming while preserving the conveniences of modernity, this transition risks trading one set of resource dependencies and environmental injustices for another. Scaling the deployment of clean energy technologies at the pace needed to avoid climate change will intensify the demand for a wide range of highly specialized materials, such as lithium, nickel, and graphite, that are sourced from around the world, with implications for workers and frontline communities. Such a transition will mean not only weaning the world off fossil fuels; it will also mean scaling up the mines, supply chains, and recycling infrastructure important to a clean energy future in ways that are both just and sustainable. Accomplishing those goals is going to require thinking about a clean energy future from the ground up.

TELLING STORIES ABOUT BATTERIES

As a historian, what strikes me about a battery-powered future is how little of this is actually new. Since the start of the twentieth century, batteries have played a little-appreciated yet pivotal role in enabling the systems of transportation, communication, and electrification that reshaped the modern world. Although the batteries of the past stored far less energy and performed less well than those anticipated for the future, the historical parallels are surprising. Many of the earliest automobiles were electric. By the 1910s, the most capable electric cars could travel 80 miles on a charge and travel at speeds up to 20 miles per hour.[16] Batteries also played an important role in enabling early telegraphs and telephone systems. Although many people fret about keeping smartphones charged today, early telephone systems depended on batteries too. In the 1910s, the forerunners of Energizer and Duracell credited the telephone industry with creating the first big market for disposable batteries. By the 1930s, as such uses for batteries disappeared, radios became even more important markets for battery manufacturers. Batteries also played an important role in the development of the electrical grid. On early electrical grids, batteries served as a buffer, ensuring a constant supply of electricity and a steady voltage. Rooms full of batteries helped to meet periods of peak power demand and provided backup power in case of emergencies.

Not surprisingly, most of the public attention devoted to batteries today focuses on prospects for the future, not technologies of the past. That public attention generally follows two related story lines. The first is a Promethean narrative focused on the next great battery breakthrough. The second sees batteries as the Achilles' heel of a clean energy future. Both story lines raise questions that this book addresses, as it considers how the history of batteries can help us think about prospects and challenges for a more sustainable and just clean energy future.

The Promethean story line is defined by breathless reporting focused on the next technological breakthrough: aluminum graphite batteries for phones that can be charged in a minute, liquid batteries that use earth-abundant materials to provide inexpensive backup power for entire cities, and lithium-air batteries that allow electric cars to match the refueling rate and range of gasoline cars. Although such breakthroughs may come, no such breakthrough technologies have yet transformed the market. Instead, the history of batteries has long been characterized by incremental advances. Batteries have never approached the dizzying pace of progress that has distinguished the modern electronics industry since the 1980s. Although computers, hard drives, and digital cameras all improved by leaps and bounds, even as prices plummeted, the same cannot be said of batteries. Advances in battery manufacturing led to substantial improvements in performance and price—indeed, lithium-ion batteries outperform older nickel-cadmium batteries on every metric—but revolutionary changes in capacity, charging rates, weight, or cost will likely require new battery chemistries or alternative technologies. That has meant that improvements in batteries have been slower than the rest of the electronics industry.[17]

Observers who see batteries as the Achilles' heel of a clean energy transition raise a different, but related, set of concerns. If there is no revolutionary new battery technology, can incremental improvements to existing technologies meet the demands of a clean energy future? A pressing question centers on the issue of scale. How many batteries will be needed to support a fossil fuel–free electrical grid? One study estimated that such a grid in the United States, largely dependent on wind and solar, would require 150 times more batteries than the Tesla Gigafactory can manufacture in a year.[18] If the United States electrifies 50 percent of its cars, that will require the annual production of another four hundred Tesla Gigafactories. Even if that production is spread out over a decade, it would require more than fifty Gigafactories operating at full capacity every year. Yet that level of production would only meet the United States', not the world's, needs.

The scale of battery production needed to support a global transition to a clean energy future is extraordinarily large. Other strategies are likely to help: demand response (such as not running appliances in late afternoon) could lower peak power demand; a shift to electrified public transit or bikes could lessen the need for individual electric cars; modernizing the electrical grid to more efficiently shuttle electricity from place to place could reduce storage demands; electric vehicles might serve double duty and provide energy storage for the electrical grid at times of peak demand; and other technologies, such as storing hydrogen or heat, might offer alternatives. But, especially in the coming decades, when change must come quickly to avoid the worst effects of climate change, it is likely that the lithium-ion battery, brought to market in the early 1990s, will do the heavy lifting in a clean energy transition.

Producing batteries at such scale requires more than just manufacturing capacity; it will also require a massive investment of energy and mobilization of materials. This is true of a transition toward a clean energy future more generally. By 2050, the energy-relevant minerals needed to largely phase out fossil fuels and scale up renewable energy deployment could approach 200 million tons per year.[19] Scaling up production of lithium-ion batteries to electrify half the world's passenger cars by 2050 will drive demand far above 2019 production levels for lithium, cobalt, and nickel—demand could outstrip supply of lithium and cobalt as early as 2025.[20] In short, decarbonizing the economy may curb society's demand for fossil fuels, but it is going to mean ramping up use of a different set of nonrenewable materials and chemicals. The geoscientist Olivier Vidal and colleagues describe this as "metal-energy dependence."[21]

Yet, for all the discussion of the importance of decarbonizing the global economy and transitioning to a clean energy future, there has been relatively little discussion of the potential material consequences of this transition. How will scaling up the production of such materials affect local communities that are the sites of mines and refineries? In the rush to scale up production, how is the safety and health of workers being protected? Will handling these materials be outsourced to countries with weaker environmental- and worker-safety policies? And is it really possible to "close the loop" on these new technologies, recovering materials from old batteries and recycling them into a new generation of clean energy technologies?

Just as the climate challenge requires careful forethought and aggressive action, so too will building the mines, the supply chains, and the recycling infrastructure needed to enable a sustainable and just clean energy future.

ENERGY QUALITY, THE (IN)VISIBILITY OF ENERGY, A CULTURE OF MOBILITY, AND A JUST TRANSITION

In investigating the environmental history of batteries, this book engages three ideas that challenge the usual thinking about the past and future of energy and environmental sustainability: the importance of energy quality; the ways in which batteries make energy visible; and how batteries underpin a broader culture of mobility. In doing so, it aligns with calls for a "just transition," which refuse to reduce the climate crisis to an energy problem or carbon crisis, and instead elevate environmental justice as a centerpiece of a clean energy transition.

Historians have long emphasized the central role of energy in understanding modern history. As John McNeill explains in *Something New under the Sun*, "no other century—no millennium—in human history can compare with the twentieth century for its growth in energy use."[22] In this traditional story, which centers on the rise of fossil fuels, nuclear power, hydroelectricity, and the electrical grid, batteries are, at best, a footnote. Even if batteries are trivial in terms of the amount of energy they deliver, batteries have occupied an outsized, but little-appreciated, role in modern energy systems for a different reason: energy quality.[23]

This is important to understanding all energy sources. It was not just the abundance of fossil fuels that made coal and oil so important to human history but also the density, storability, portability, and relatively low cost of fossil fuels. Considering energy quality also helps explain why renewable energy sources, such as solar and wind, struggled to compete with fossil fuels in the twentieth century. Although the quantities of energy available through wind and sunshine on Earth far exceed the quantity of energy stored in fossil fuels, the distinguishing qualities of wind and solar are their intermittency, lack of density, and lack of storability—all of which put them at a disadvantage compared to fossil fuels or hydroelectricity. It is these same qualities that help explain why batteries, despite their limitations, have been so important in the past and will continue to be in the future. Batteries are relatively unique in their portability, storability, and ability to supply nearly instantaneous electricity on demand—indeed, it is these qualities that make them a key component of a clean energy future.

Even if batteries have drawn little attention from historians, batteries have occupied an important place in how most people think about energy. Inevitably one of the first questions I get asked about my research is the practical one: What should we do with our old batteries? Can we recycle them or should we toss them into the trash? As I explain in the following chapters, the answer depends on the type of battery. But the question itself offers insight into the place that batteries occupy in the cultural history of energy.[24] One of the themes that emerges from histories of American energy use is that as the availability and reliability of energy improved, attention to it diminished. This is one of the defining qualities of modernity: the intensification and the abstraction of energy consumption.[25] Since the twentieth century, when electricity became available at the flip of a switch and gasoline could be pumped on demand at the gas station, energy could often be taken for granted.[26]

Arguably, this is one of the great accomplishments of the energy industry. Electric utilities have had a "long and successful" history of making "its product largely invisible, both in its manufacture and physical manifestation."[27] But batteries have always demanded consumers' attention. For much of the twentieth century, automobile drivers had to be sure to top off the electrolyte in their car's starter battery. In the era of the Sony Walkman, people fretted over the disposal of mercury-containing AA batteries, which many rightly regarded as small bundles of toxic waste. And in the age of the smartphone, keeping lithium-ion batteries charged has become the nag of modernity. In a world where energy seems at once ubiquitous and largely invisible, batteries demand a great deal more attention—whether it is replacing them, charging them, or deciding how to dispose of them—than almost any other component of the modern energy system.

When energy has been a pressing public issue, those discussions have most often centered around mobility—moving people and goods from place to place. For example, most portrayals of the 1970s oil crises start with a photograph of a long line of cars waiting to fill up at a gas station. More broadly, environmental scholars and environmentalists have long been concerned with the consequences of mobility: indeed, among the most energy-intensive and polluting activities that humans undertake is moving themselves and goods from place to place. That story often begins with the Model T and ends with global warming. Some have even argued that we should put oil at the center of modern historical analysis; no doubt we can learn much by doing so.[28] What environmental scholars have given less attention to, however, is the environmental consequences of another form of mobility: moving information from place to place. Less attention has been given to the environmental consequences of radios, telephones, and other forms of modern communication. Indeed, those technologies—then and today—often seem ethereal, transcending the bounds of the physical world.

Yet, batteries highlight the ways in which all of these forms of mobility—moving people, things, and information—intensified the demands on the material world. On this point, I take a cue from the field of mobility studies.

The mobility paradigm focuses on the interconnected mobilities of people, information, capital, goods, and services. As the sociologist John Urry has argued, "the irreversible consequences of the last 'mobility century' and the extraordinary 'digitization' of life have left an awesome interdependent legacy."[29] The "new mobilities paradigm" aims to understand how changes in mobility, broadly conceived, have reconfigured modern social life, civil society, and governance. Considering the history of batteries makes clear that this "interdependent legacy" depended not just on energy and fossil fuel extraction but the intensified use of metals and chemicals too.

This history of batteries taps into an emerging thread in discussions about a "just transition" too. Since the early 2000s, for those on the front lines of progressive climate policy, a just transition has become an organizing concept in climate activism. It emphasizes the need to ensure that the workers and communities whose livelihoods depend upon the coal, oil, gas, and related industries are not left behind as the world weans itself off fossil fuels. It criticizes the fossil fuel economy as one "based on extracting resources from a finite ecosystem faster than the capacity of the system to regenerate" and makes clear the disproportionate impacts a fossil fuel–powered extractive economy has had on indigenous peoples, communities of color, and the poor. Proponents of a "just transition" urge us to move from an extractive economy to a regenerative economy that empowers local communities, workers, and the "countless others who have been harmed by the extractive economy."[30] Central to this vision is a shift from fossil fuels to a new "energy democracy" that is appropriately scaled, empowers local communities, and relies upon renewable energy.

Climate justice advocates have begun to consider how building a clean energy economy for the twenty-first century risks reproducing the injustices of the twentieth century.[31] Indeed, winding down the use of fossil fuels is going to mean winding up the use of a wide range of other resources important to a clean energy future at a pace that exceeds the growth of the twentieth-century fossil fuel economy. As the authors of *A Planet to Win: Why We Need a Green New Deal* have argued, we are at a fork in the road on the way to a clean energy future. It could take the form of "solar-powered capitalism," reproducing the problems of the fossil fuel–powered twentieth century, driven by "a whole new set of opportunities for profit and pillage," or it could become a "historic opportunity to remake global power structures and our relationship to the natural world" that puts the workers, communities, and the environment ahead of "corporations and financiers."[32] These are questions to which ecomodernists, with their unalloyed faith in the promise of technological advance, have given little consideration. Considering the history of batteries offers a sharp reminder that, just as climate justice advocates have emphasized, solving the equation of a clean energy future will be about far more than just zeroing out carbon.

THREE STORIES: LEAD-ACID BATTERIES, AA BATTERIES, AND LITHIUM-ION BATTERIES

The idea for this book began in the classroom. I teach in an environmental studies department where alarm over climate change and its consequences for the planet and its most vulnerable residents, both human and nonhuman, has increasingly driven my students' academic work and their activism. They are tuned to the guidance of the Intergovernmental Panel on Climate Change, which urged the world to reduce greenhouse gas emissions by nearly half by 2030 and eliminate fossil fuel consumption almost entirely by 2050.[33] In their view, addressing climate change is fundamentally an issue of social justice, which poses the greatest threat to disadvantaged communities today and in the future. My students have expressed their admiration for leaders such as Greta Thunberg and Alexandria Ocasio-Cortez, who have transformed climate activism. On campus, they have successfully pressed the college to divest its endowment from fossil fuels and adopt plans for achieving carbon neutrality. Off campus, they have championed policy goals aligned with the Green New Deal.

Despite their concerns about climate change, interest in energy systems, and concerns for social justice, if I ask my students how batteries fit into the clean energy equation, they often have little to say. And if I ask whether they have thoughts about the scale of materials production necessary to transition to a clean energy future more broadly, that usually does little to further the discussion. Although my students depend upon batteries every day to power laptops and cell phones, and they foresee an important role for batteries in a clean energy future—powering electric cars and backing up renewable energy sources—they know little about the different types of batteries, what is in them, how they should be disposed of, or how that might change in the future. My students' uncertainty about batteries is true of the public more generally. Although there are numerous books aimed at a public audience about other energy-related topics, such as coal, oil, nuclear power, or renewable energy technologies, books about batteries are in shorter supply.[34] In short, batteries are a black box. My goal is to pry that box open and, in doing so, to think about what batteries can teach us about the material consequences of a clean energy future more broadly.

An important place to start this book, then, is by explaining what a battery is. Although there are other ways to store energy—in hydropower reservoirs and spinning flywheels, or by compressing air or heating up or cooling down water—what makes batteries unique is that they store energy in chemical form, often in portable devices, which can be converted into electricity nearly instantaneously and on demand whenever and wherever. Batteries are made up of one or more cells. Each cell contains an anode, a cathode, and an electrolyte. When connected by a circuit, a battery releases electrons from the anode (which then travel through the electrical circuit from the negative to the positive terminal), while the anode and cathode undergo complementary chemical reactions (exchanging ions through the electrolyte). As the electrons move from the anode to the cathode, they can do useful work, such as powering a lightbulb, a motor, or a microprocessor. If the battery is rechargeable, applying electrical energy reverses the chemical reaction, restoring the battery's charge. In the lingo of battery engineers, which I'll use in this book, a *primary battery* refers to a single-use battery, such as the AAA in a remote control, and a *secondary battery* is a rechargeable battery, such as the lead-acid starter battery in a gas-powered car or the lithium-ion battery in a smartphone or an electric car.

The principles of a battery are surprisingly simple. A classic demonstration is the lemon battery, in which a zinc nail serves as the anode, a copper coin as the cathode, and the acidic lemon juice as the electrolyte. When the zinc nail and copper-plated penny are inserted into the lemon and then connected by a wire, the battery has a voltage of approximately 0.9 volts. If four lemon batteries are linked together in series, they can generate approximately 3.6 volts and just enough current to light up a small LED bulb. As the zinc and copper react with the acidic lemon juice, electrons flow from the zinc nail, through the circuit, to the electron-hungry copper penny. (This is easy to do. You should try it yourself!)

Although all batteries share this basic chemistry, a battery's properties depend upon the specific combinations of materials that make up the anode, cathode, and electrolyte. The exacting qualities of those materials determine a battery's capacity, power, durability, and other properties, such as whether it is a single-use primary battery or a rechargeable secondary battery. Indeed, battery engineering is an exercise in material chemistry that turns on levels of purity measured in parts per million, the deliberate additions of beneficial impurities, and the unique physical properties of materials, both familiar and rare.

Historically, only a handful of battery chemistries have struck the right balance between reliability, performance, and economy to reach the market. But the batteries that have struck that balance have been deployed on a massive scale. The single-use AA flashlight battery is one of the most common. Over time, its chemistry has changed to maximize shelf life, reduce toxicity, and enhance performance.[35] But it is a primary battery: it cannot be recharged. Historically, the most common secondary battery has been the lead-acid automobile starter battery. It is optimized for durability, all-weather performance, and providing the short bursts of current needed to start a car. Since the early 1990s, lithium-ion batteries have increasingly met the growing demand for portable power. There are a variety of lithium-ion battery chemistries, most of which use a lithium-based cathode, a graphite anode, and an electrolyte. Although the performance of lithium-ion batteries varies depending on the chemistry, in general they are more powerful, more durable, and have a higher capacity per unit weight than lead-acid batteries. Those advantages have made them the preferred technology for portable electronics, electric cars, and, increasingly, storing energy on the electrical grid.

This book tells the story of the three battery chemistries that dominated the twentieth century—the lead-acid car battery, the disposable AA battery, and the newer lithium-ion battery. Each of these case studies is instructive, as we consider the prospects for a battery-powered clean energy future. Lead-acid batteries are the most efficiently and highly recycled product in the world. AA batteries are among the most polluting sources of electricity commonly used. And lithium-ion batteries have driven a growing resource omnivory, as they require a wider array of highly refined materials to perform. In different ways, each of these case studies offers insights into the challenges of scaling up batteries to support a clean energy future.

Telling these stories has tested my skills as both a researcher and story-teller. Archival materials often lay behind the firewall of the consumer products or global electronics industries. Although I rely on some archives, I have also drawn on a variety of less conventional sources, such as webinars, investor reports, records of trade disputes, and historical satellite imagery. Limited sources means that at times this book tends toward a history of seemingly inevitable technological advance. Considering the material complexities of batteries, and how the systems of sourcing and manufacturing them have changed over time, this narrative strategy makes these case studies easier to follow—especially for readers not steeped in the details of material science. But it is an approach that lacks the attention to contingency, agency, and context that historians of technology value. Lastly, while this book is concerned about the local impacts and injustices relevant to the history of batteries, both in the United States and abroad, this book is not written from that perspective. Instead, it is largely organized around my perspective as a consumer, both counting on and concerned about the role of clean energy technologies in propelling us toward a more sustainable future.

TOWARD INDUSTRIAL ECOLOGICAL LITERACY

The more I have thought about batteries, and their importance to a clean energy future, the more uneasy I have become about some of the basic assumptions that underlie modern environmental thought. Environmentalists have long been concerned by how modern technology and consumer culture distance humans—specifically those living in modern, urbanized societies such as the United States—from the natural world. If we push aside the thickets of green consumerism and complexities of environmental regulations, the prescriptions for righting our relationship to the natural world follow some familiar paths: choosing the natural rather than the synthetic, the local rather than the global, and material simplicity rather than plenty. Taken together, proponents of "ecological literacy" urge the world to embrace a "new vision . . . that links us to the planet in more life-centered ways."[36]

Batteries do not fit neatly into prescriptions for "ecological literacy," however. They are a form of technology that depends upon wresting raw materials from the Earth, applying immense amounts of energy to refine those materials and assemble them into batteries, often at some risk to both the environment and human health—all of which is obscured by the complex chains of materials and manufacturing that make them possible. In this respect, batteries are representative of many of the technologies needed to advance a more sustainable future, including solar panels, wind turbines, and electric motors.

In this book, I take batteries as the starting point for an investigation into the paradoxes of modern technology and their social and environmental consequences. In doing so, I turn the original proposition—that technology distances humans from nature—on its head. I argue that some of our most consequential and meaningful interactions with nature are in the built environment: the infrastructure that undergirds our communication, transportation, and energy systems, the networks of production and consumption that stock our shelves with innumerable technological goods, and the commodity chains that supply us with the materials that make all of this possible. Environmentalists have long argued for "ecological literacy" as a key tool in fostering the changes needed to address the modern sustainability crisis: ecological literacy is meant to educate people about the complex natural systems that make life on Earth possible and how those principles can inform a more sustainable future. I argue that we need to give equal attention to a form of industrial ecological literacy that can reveal how these technologies entangle us in the material world, with vast social and environmental consequences.

Leaning into industrial ecological literacy reminds us that a clean energy future is not just about what it will create—resiliency, sustainability, or energy democracy—but what it will consume: energy, resources, and materials. Taking that into account reframes what is required to address climate change at a pace commensurate with the imperatives of climate science and with attention to the issues of social justice at the heart of a just transition.

To advance those goals, this book concludes by outlining four broad priorities to help chart the future of clean energy policy: (1) the importance of leveraging the power of the government not only to hasten the transition to a clean energy future but to secure the resources needed to enable that transition; (2) the need to implement strategies to ensure that the global trade in minerals and materials—which will be essential to scaling up clean energy technologies, including batteries—is both sustainable and just; (3) the imperative to expand domestic minerals production in the United States to support a clean energy transition, which will require both reforming existing mining practices and setting goals for expanding minerals development and processing domestically; and (4) ensuring, despite prospects for a circular economy, that an emphasis on recycling does not distract from the near-term imperative to massively scale up the production of energy relevant minerals domestically and worldwide. Each of these strategies is going to be important to building a clean energy future, from the ground up.

1

LEAD-ACID BATTERIES AND A CULTURE OF MOBILITY

IN DECEMBER 1996, GENERAL MOTORS RELEASED THE EV1, A NEW electric car with swift acceleration, zero maintenance, keyless entry, and a futuristic digital display that consolidated the speedometer and other instrumentation in the center of the dashboard, leaving the driver with an undistracted view of the road. *Motor Trend* described it as a "modern-day X-wing fighter" that rendered obsolete any "thoughts of electric cars as having feeble golf-cart-like performance."[1] It was the first electric car that General Motors had put into production since the 1910s.

The 2006 documentary *Who Killed the Electric Car?* tells the story of the rise and fall of the ill-fated EV1. Despite consumer enthusiasm, GM manufactured only 1,117 vehicles, leased them primarily in California, and discontinued the car due to what it described as low demand.[2] The documentary contrasts the EV1's enthusiastic lessees, who included celebrities such as Tom Hanks, with the regulators, oil companies, and General Motors itself, which all seemed determined to sabotage what could have been a bright future for electric vehicles.

But, in retrospect, the original EV1 did have a fatal flaw: its batteries.

When the $34,000 EV1 hit the market in 1996, its range was 70 to 90 miles on a good day. But it charged slowly, the early battery packs proved unreliable, and the car's performance was fickle: cold or hot weather sharply reduced its range. The EV1 may have offered a glimpse into an electric-car future, but it depended upon a battery chemistry first commercialized in the late nineteenth century: the lead-acid battery.

Based on the rise and fall of the EV1, anyone looking ahead to a more sustainable future would likely conclude that lead-acid batteries represent a dead end: too heavy, too inefficient, and too toxic to play a pivotal role in the transition to a clean energy future. Even GM thought so. In the late 1990s, it swapped out the lead-acid batteries in the first generation EV1 for better-performing nickel–metal hydride batteries—similar to those that would propel the Toyota Prius to success in the 2000s.

For all of these reasons, the history of the lead-acid battery might seem like a surprising place to begin this book. But to set up this chapter, let me offer four reasons why the history of the lead-acid battery—which, to paraphrase William Faulkner's quip, is not even past—should be at the forefront of our thinking about batteries, materiality, and prospects for a transition to a clean energy future.

First, lead-acid batteries are ubiquitous in the built environment. Today, every vehicle, including every car, truck, motorcycle, and motorboat depends on a lead-acid battery for starting, lighting, and ignition. The battery industry has manufactured lead-acid batteries on the scale of Gigafactories for decades. Even though the use of lithium-ion batteries is growing rapidly, lead-acid batteries still account for the majority of the energy-storage market. Thus, if we want an approximation of the scale of lithium-ion battery production necessary to support a clean energy future, the lead-acid battery offers an important and cautionary tale.

Second, lead-acid batteries are the most highly recycled product in the world. Since the 1990s, more than 97 percent of lead-acid batteries have been recycled in the United States. Critics often point out that recycling is often just a form of downcycling, yielding lower-quality and less-valuable secondary materials. Not so with lead-acid batteries: since the 1970s, integrated battery breaking and smelting operations, as they are known, have recovered lead, sulfuric acid, and plastic casings with great efficiency. Most of those materials are used in new batteries, offering a model for a circular economy that turns waste into high-quality new products. History suggests those claims deserve scrutiny.

Third, even as the odds of a lead-acid battery getting recycled neared 100 percent in the United States, that statistic is troubling in the context of a system that processes over one million tons of spent batteries per year. If 1 percent of lead-acid batteries go unrecycled, that means more than ten thousand tons of lead go unaccounted for each year, in the form of batteries that get landfilled, abandoned in garages, or covertly shipped overseas. Although the use and reuse of lead in batteries is measured by the ton, lead contamination measured by the part per million in the air, soil, and, ultimately, in human bodies poses a grave risk to human health. As has been the case with most environmental health threats, workers, the poor, and people of color have often disproportionately borne those consequences. Considering the history of the lead-acid battery industry highlights successes in recognizing and managing those risks, but also devastating failures.

And fourth, historically, lead-acid batteries were largely a domestic industry, mined, manufactured, used, and recycled in the United States. The acceleration of globalization and shifts in trade and environmental laws since the 1970s has changed that: the burdens of sourcing and recycling lead-acid batteries have increasingly been shifted outside of the United States to other countries with lesser protections for public health and the environment.

This chapter offers a short account of the long history of the lead-acid battery. It is the story of the most ubiquitous, toxic, highly recycled, and inherently unsustainable battery technology. This chapter explains how all of those things can be true, simultaneously.

THE USES OF LEAD-ACID BATTERIES

Today, proponents of a clean energy transition envision a future in which batteries play a new role in powering cars, enabling communications systems, providing backup power, and meeting periods of peak energy demand on the electrical grid. Although such visions mark a sharp break with the fossil-fueled past, none of those applications is entirely new. At the start of the twentieth century, to the extent that people counted on telephones, electrical grids, and early electric cars as ready and reliable devices, they were counting indirectly on the growing role of lead-acid storage batteries to deliver electricity on demand.

The French chemist Gaston Planté is credited with inventing the lead-acid storage battery in 1859. The electrical inventors Elihu Thomson and Edwin Houston foresaw its potential applications in telegraphy, lighting systems, and other electrical systems in the 1870s and 1880s. By the early twentieth century, what had been largely a "laboratory curiosity" in the late nineteenth century came to play a key role in supporting emerging systems of electrical distribution, transportation, and communication.[3]

Consider this abridged list of the many uses of lead-acid batteries in the 1910s: Storage batteries paired with generators provided small-scale electrical grids for farms wanting electric doorbells, lights, and pumps. Railroads depended on storage batteries to power signals, safety devices, and railcar lighting. Early electrical companies relied upon storage batteries for battery backup and to meet short periods of peak electricity demand.[4] Although early telephones depended upon dry-cell batteries, as we'll see in chapter 3, when telephone companies transitioned to centralized switching operations, storage batteries played a key role regulating the electrical supply. As one engineer explained: "The use of the storage battery in the telephone and telegraph fields has gone hand in hand with the building of bigger and better telephone exchanges and systems. No other form of power supply can deliver as steady an electric current without any fluctuation in current or voltage as the storage battery."[5]

Yet, no one foresaw the application that would drive the growth of the lead-acid battery industry: gasoline-powered automobiles. To the extent that battery manufacturers focused on transportation at the start of the twentieth century, they focused on electric cars as a competitor with early gasoline vehicles. And, at first, that strategy worked. Until the 1910s, there were as many electric cars on the road as gasoline cars.[6] The Mark III was a popular early electric car, with a maximum speed of 12 miles per hour and a range of 35 miles—competitive specifications in the infancy of the auto industry (before the advent of high-speed roads and long-distance travel). As the historian Virginia Scharff explains, manufacturers often pitched electric cars to women, who they believed would appreciate these vehicles as "simple, comfortable, clean and quiet," even if they were "short on power and restricted in range."[7]

The EV1 was not the first time General Motors had a hand in killing the electric car. It did so in the 1910s too. In 1911, an independent inventor, Charles Kettering, invented an electric starter for gas-powered cars. It incorporated a storage battery and a small electric starting motor, which made it possible to start a gas-powered car with the push of a button or, later, the turn of a key. Cadillac Motor Company, which was already a division of General Motors, introduced the electric starter in its 1912 model year. It soon became standard equipment on all cars, making them as safe and convenient as electrics.[8]

SEVENTY=FIVE MILES
ON ONE CHARGE

is the performance we claim for our No. 1,000 Electric, at a speed of 13 miles per hour—at 20 miles per hour it will go 40 miles on one charge.

This unusually high mileage is secured by a scientifically correct electrical installation, and a nearly frictionless running gear.

All of the many good points of this car are fully described in the catalog—may we send it?

THE COLUMBUS BUGGY CO.
COLUMBUS, OHIO

FIGURE 1.1 Early electric cars competed with gasoline and steam cars. Manufacturers emphasized their convenience, range, and speed. *The Horseless Age,* November 7, 1906. Courtesy of chuckstoyland.com.

As a result, the most widespread application of lead-acid batteries was not for powering cars; it was for starting cars and powering their lights and, eventually, their turn signals, windshield wipers, and radios. Broadly speaking, lead-acid batteries became a key enabling technology for an emerging culture of mobility in the twentieth century. As one treatise on the battery surmised in 1917, "A rather flexible imagination is needed to consider that the cheerfully lighted farm home; the lifting of a massive drawbridge; the roaring start of a powerful automobile or hydroaeroplane engine and the noiseless movement of the electric automobile are all accomplished by the same agency."[9]

As sales of automobiles rose in the 1920s, so too did sales of lead-acid starter batteries: fifteen million in 1930, thirty-three million in 1950, and more than one hundred million in 2010 in the United States. Although automobile starter batteries dominated the use of the lead-acid battery during the twentieth century, other uses of lead-acid batteries did not disappear. In fact, starting in the late 1980s, industrial applications of lead-acid batteries began to accelerate, with the rise of cellular telephone networks and the internet. Lead-acid batteries provided battery backup for a growing list of critical communication systems, ranging from emergency call centers to cell phone towers to cloud data servers to home internet routers. After the 1980s, this was the fastest-growing market for lead-acid batteries.[10] Meeting this growing demand for lead-acid batteries required not only more raw materials but, as would often be the case for other battery technologies, purer and higher-quality materials, the production of which often posed threats to the environment and human health.

WHAT MAKES A LEAD-ACID BATTERY WORK?

Remarkably, the basic chemistry of lead-acid batteries has remained relatively unchanged since the early twentieth century. When you start a gas-powered car today, a surge of electrons flows out of the starter battery's negative terminal, through the car's starter motor, to reach the battery's electron-hungry positive terminal—just as it did in the 1910s or 1920s. What has changed is the reliability and longevity of lead-acid batteries. In the early twentieth century, batteries were notoriously finicky: unreliable, short-lived, and temperamental.

Manufacturers' success in overcoming those challenges depended upon slight changes in materials and amendments that improved battery performance. Consider the early history of the Willard Storage Battery Company. Its Cleveland, Ohio, factory manufactured six thousand storage batteries per day in the 1920s. The general form of the lead-acid automobile battery was largely standardized by the 1930s. Three 2-volt cells were wired in series to supply 6 volts to the car's starter motor and accessories. Each cell was made up of positive and negative lead plates, each plate separated by a thin piece of wood, and packaged in a wooden case filled with sulfuric acid.[11]

The Willard Company promised that every battery "has been carefully watched, checked, and rechecked" in its laboratories to ensure "that each item conforms to the high standards laid down by the Willard Engineers."[12] Although we cannot trace the paths of the materials that supplied the Willard Company fully or with precision, over time the catchment area supplying its raw materials—sulfuric acid, antimony, wooden separators, and other additives—became increasingly extensive, spanning the United States and, in some cases, the world.

Sulfuric acid was essential to lead-acid batteries. When sulfuric acid is dissolved into water, it produces protons and ions containing both sulfur and oxygen, which are ready to react with the electrodes in a battery. Thus, the voltage of a lead-acid battery is as much a product of the properties of sulfuric acid as it is of the lead electrodes. The battery industry was just one of many major industries that relied upon sulfuric acid. It played a key role in the production of steel, petroleum, fertilizers, and paper. As early as the 1920s, economists viewed the consumption of sulfuric acid as an indicator of a nation's degree of industrial progress.

Until the start of the twentieth century, however, the United States depended almost entirely on imported sulfur, much of which came from Sicily. In 1900, Italy exported 162,505 tons of sulfur to the United States, accounting for nearly one-half of American consumption.[13] The Sicilian mines were notorious for grueling working conditions and child labor. The Black educator and political leader Booker T. Washington reported on his visit to the Sicilian mines in 1911. He likened the working conditions of the *carusi*, or enslaved children, to "the cruelties of Negro slavery" and the mines to a living "hell."[14]

In the 1880s, early petroleum exploration along the Gulf coast revealed extensive sulfur deposits south of New Orleans. The problem was how to mine it. Unlike in Sicily, where sulfur ore could be mined underground, the Louisiana sulfur deposits were located in a coastal area where sand, unstable sediments, and water made conventional mining practices impossible. A string of companies tried and failed to bring Louisiana's sulfur into commercial production in the nineteenth century, leaving the United States largely dependent on Italy.[15]

One of the most curious, but little-known, stories in American mining history is the discovery of the Frasch process for sulfur recovery. After two years of experimentation, the Union Sulphur Company successfully implemented a strategy pioneered by a French petroleum engineer, Herman Frasch. Instead of laboriously transporting sulfur ore to the surface, as they did in Sicily, Frasch's strategy took advantage of sulfur's low melting point (240°F) and its density (it was twice as dense as water). It was an ingenious strategy that required pumping superheated water into the sulfur formation six hundred feet below the surface, melting the sulfur, which then sank to the bottom of the well, where it could be forced to the surface. As one mining historian wrote, Frasch "succeeded in penetrating the strongbox of nature and extracting the magic mineral."[16]

The Frasch process transformed American industry in the early twentieth century. Louisiana's supply of sulfur was seemingly inexhaustible.[17] It was energy-intensive but economical too. By relying on fossil fuels to heat the water to melt the sulfur, the Frasch strategy allowed four hundred miners in Louisiana to produce as much sulfur in a day as did twenty-one thousand miners in Sicily and with far less risk to the miners. Between 1910 and 1920, US production of sulfur grew sixfold to 1.2 million tons per year, more than 99 percent of which was accounted for by the Frasch process in mines in Louisiana and, starting in 1913, Texas.[18] The subsequent collapse of the Sicilian sulfur industry contributed to the surge of Sicilian emigration to the United States in the early twentieth century.

For a starter battery to be usable, it had to deliver the high current necessary to start a car over and over again. To deliver a high current, a car battery requires thin electrode plates that maximize the surface area exposed to the electrolyte. But making thin plates was difficult for two reasons: first, molten lead did not flow easily through thin molds because of its viscosity, and, second, once cast, thin plates were prone to crumble after extended use. Starting in the late nineteenth century, battery manufacturers began to use a lead-antimony alloy, which resulted in thinner and stronger plates. As one industry journal explained, "Antimony serves chiefly as a backbone, so to speak, for other metals."[19]

The use of antimony imposed another limitation on early storage batteries. It increased the rate at which batteries released hydrogen from the electrolyte as they cycled. That may sound like a technical issue, but it had very practical consequences for car drivers: lead-acid batteries had to be regularly topped off with water. Not until the 1970s did the industry introduce a maintenance-free lead-acid battery.

Building a battery that could fit under the hood of a car required packing the thin positive and negative plates together more tightly than in the stationary batteries used for telegraphs or electrical systems. One of the problems in lead-acid batteries was the tendency of thin plates to deform over time and short-circuit the battery. This became more likely as plates became thinner and were packed more closely together. Separators placed between the plates prevented such short circuits, while also being permeable to the electrolyte that allowed for the flow of ions during discharging and charging cycles.

Industry experimented with many different separator materials, including various types of wood, such as cherry, poplar, pine, and fir. By the 1930s, Port Orford white cedar, which is found along the southern coast of Oregon and northern coast of California, had become a preferred wood for many battery manufacturers because of its porosity.[20] The Evans Products Company based in Coos Bay, Oregon, dominated the separator industry, delivering slices of white cedar 5/64 inches thick, grooved with parallel ribs, and kiln-dried to meet the needs of battery manufacturers nationwide.[21]

The forest-products industry proved important to the storage-battery industry in another way. The potential of sawdust to improve battery performance became apparent when battery manufacturers began experimenting with rubber separators in place of wood separators. The technique, pursued by the Willard Storage Battery Company, revealed that the negative battery plates in batteries without wooden separators lost capacity more quickly than expected. That suggested that the presence of the wooden separators was contributing to battery performance in an unappreciated way.

By the 1920s, engineers realized that a small amount of sawdust, added to the negative battery plates, improved performance by increasing their porosity.[22] By the 1930s, manufacturers had further refined the technique, identifying the key role of lignin in the performance of negative battery plates. That realization gave new value to black liquor, a lignin-rich waste product resulting from paper mills. The end result was lead-acid batteries with 10 to 20 percent better performance—as measured by current delivered—in cold-weather conditions than battery plates without the additives.[23]

The 1928 Willard Storage Battery owner's manual is of a particular vintage in American industrial history: it was as much an industrial biography detailing the manufacture of a battery as it was an instruction booklet on how to operate it. This is worth noting in comparison to today's batteries, which are literally black boxes. But in the 1920s, Willard's manual explained with much pride the different purposes that lead, lead oxides, sulfuric acid, and separators all played in a battery that measured up to the Willard Company's standards.[24]

The overarching theme was control. Careful sampling of raw materials, the precise regulation of each stage of manufacturing, and the rigorous testing of final batteries were all essential. And by all accounts, Willard and many other companies succeeded in this respect, drawing on a wide range of raw materials to deliver millions of batteries to market every year. Both in the early twentieth century and today, batteries often appear as functional black boxes, posing no immediate threats to human health or the environment while in use.

Yet, neither Willard nor any other battery manufacturer was ever fully in control of the materials or the processes upon which their batteries depended. Although batteries are often safe while in use, nearly every other step in the life cycle of a battery, from sourcing the raw materials to managing them at end of life, poses risks to the environment and human health that must be managed. Those risks were especially high in the case of lead-acid batteries.

THE HUMAN TOLL OF MAKING LEAD-ACID BATTERIES

In the early twentieth century, as concerns about the threat of industrial toxins began to emerge, the lead industry became the focal point of occupational health studies and industrial hygiene in the United States.[25] At the time, lead had already become common in the built environment: it was in pipes that carried water, it sheathed electrical cables, it was in paint, it was sprayed on crops as an insecticide, and it was, of course, in batteries under the hoods of automobiles. Few people raised concerns about lead pollution from factories or farms. Instead, concerns about lead exposure centered on workers and the hazards of the job. As a result, some of the most detailed explanations of how lead-acid batteries were made come from industrial hygienists, who studied working conditions in factories.

Here are the basic steps in manufacturing a lead-acid battery: First, workers melted lead and small amounts of antimony in kettles to prepare a lead-antimony alloy that they poured into molds to form the grids of battery plates. Second, workers mixed a paste of lead oxides by combining fine powdered lead oxide with a dilute sulfuric acid. This was the most hazardous step; even in factories with ventilation, lead dust often caked the rooms and workers. Third, workers then applied the paste to the plates, filling the interstices in the battery-plate grids. Fourth, workers moved plates to drying racks or rooms, where the lead paste set. Fifth, after trimming the plates, workers formed the plates by exposing them to an electrical current in an electrolyte bath to form negative plates and positive plates. Finally, workers assembled the battery plates in boxes, using lead solder to connect the alternating positive and negative plates, which were separated by wooden separators. Workers then added terminals, filled the battery with dilute sulfuric acid, and charged it in preparation for shipment.[26]

Every step created opportunities for lead to move into workers' bodies, most often through what they ate, drank, or, most problematically, breathed.

But these risks were unclear to early industrial hygienists. Even though physicians and plant managers had some awareness of the hazards lead posed in the early twentieth century, many blamed the high levels of plumbism, or lead poisoning, on workers who, managers believed, had weak constitutions or practiced poor hygiene. In 1914, an industry advisory group published seemingly sensible guidelines for storage-battery workers that included washing hands, trimming facial hair, changing clothes after work, and wearing protective gear such as hats and goggles.[27] In their view, worker safety was a matter of personal hygiene more so than factory conditions.

Alice Hamilton, a pioneering industrial hygienist, government reformer, and Harvard professor, shone a light on the threat of occupational health hazards in the early twentieth century. The historian Christian Warren emphasizes the importance of her legacy in *Brush with Death*, his social history of lead poisoning: "It does not exaggerate Hamilton's impact to say that she helped change American medicine's attitude toward occupational hazards from blind acceptance of the ways things were to open-eyed determination to see what could be done."[28] In the 1910s, she undertook extensive site visits and surveys at lead smelters, refineries, and manufacturing firms around the United States, including storage-battery manufacturers. Her surveys, which were largely based on her observations of working conditions, interviews with workers, and rudimentary biological analyses, revealed high levels of lead poisoning in the storage-battery industry.

Hamilton drew a contrast between older, often smaller, factories and newer, often larger, factories that had better provisions for worker safety. At many of the older factories she visited, she documented crude industrial practices, where poor ventilation, high levels of dust, infrequent cleanings, and a lack of facilities for workers (such as lockers and showers) meant high levels of lead exposure. No matter how often workers in such factories washed their hands or how neatly they trimmed their beards, they faced a high risk of lead poisoning.

Although she did not label it this way, Hamilton's research detailed the environmental injustice in how the risks of manufacturing lead-acid batteries were allocated. Based on her research, workers mixing the dusty oxides for application to battery plates suffered lead poisoning at twice the rate of other factory workers.[29] Later surveys revealed the work was organized by race and ethnicity in some factories: Black people and recent immigrants were disproportionately employed in the mixing rooms.[30] Such occupational health research demonstrated that the issue was not workers' hygiene, race, or their ethnicity that put them at higher risk for plumbism; it was where they worked in the factory and the working conditions they faced.

Hamilton proved an effective advocate for industrial reform, often working in cooperation with the lead industry, which gave her access to their factories and funded some of her research. Hamilton's primary recommendations for improving industrial hygiene focused on improving working conditions and better monitoring worker health. In making that case, Hamilton highlighted the benefits of newer, larger factories—such as the one Willard operated outside Cleveland—which had better ventilation systems; separate rooms for each stage of manufacturing; lockers, wash stations, and lunchrooms for employees; and on-site company physicians.

If Alice Hamilton starts to sound like a heroine in the history of lead industry, then Robert Kehoe fills the role of villain. Kehoe, another industrial hygienist, played a key role in setting the agenda for studies of lead toxicology and downplaying the possibility that lead posed a threat to public health, especially outside the workplace. He endeared himself to the lead industry in the 1920s because he played a key role in arguing that adding lead to gasoline as an antiknock agent posed no threat to public health. By 1930, Kehoe had become the nation's leading lead toxicologist, operating out of the industry-funded Kettering Laboratories at the University of Cincinnati.[31]

The historian Christian Warren has described Kehoe's research findings as the Kehoe-Kettering paradigm of lead toxicology. By the 1930s, Kehoe's research suggested that lead was naturally occurring in the environment, that everyone was exposed to and absorbed some lead, and that the human body was capable of protecting itself by excreting lead. Based on these findings, Kehoe concluded that there was a well-defined "threshold" for human lead exposure. As long as exposure remained below that level, lead posed no risk to human health.

Between the 1930s and 1960s, Kehoe's laboratory assembled a voluminous body of research that highlighted the occupational risks of lead exposure but downplayed the public-health risks from leaded gasoline, leaded paints, or factories. Kehoe remained a strong proponent of industrial hygiene practices.[32] Based on his observations of industry workers, Kehoe identified a blood lead level of 80μg/dL (0.8 parts per million) as the "threshold" below which he had never observed clinical symptoms of lead poisoning. He later argued that same threshold applied to the general public, including children.

By the mid-twentieth century, the lead industry prided itself on its occupational health practices. A 1943 report published by the Industrial Hygiene

Section of the American Public Health Association emphasized the decline in both fatal and nonfatal forms of lead poisoning over the previous twenty years, while highlighting the need for continued vigilance: "Experience has shown that adequate control of potentially dangerous types of lead exposure depends more upon the proper design of a plant and its equipment than any other factor."[33]

As long as the Kehoe-Kettering paradigm held and industry remained committed to good industrial hygiene, the industry viewed lead exposure a matter of concern in the workplace, not a matter of public health. But in the 1960s, in the aftermath of Rachel Carson's *Silent Spring*, which raised concerns about the consequences of industrial toxins, new research on the distribution of lead pollution and growing concerns about the consequences of low levels of lead exposure undermined the Kehoe-Kettering paradigm. The subsequent policy shifts upended the entire lead industry.

THE PRIMARY AND SECONDARY LEAD INDUSTRIES

On the cusp of battery-powered world, in which higher-capacity lithium-ion batteries play a key role in powering vehicles and stabilizing the electrical grid and proponents of a "circular economy" emphasize the possibilities for urban mining, the lead-acid battery industry sets a high standard for battery recycling. Today, nearly every single spent lead-acid battery gets recycled into new batteries. And since the mid-twentieth century, the most important source of lead for new batteries has not been a geological mine; it has been spent lead-acid batteries. Yet, the history of the lead-acid battery highlights both the potential and hazards of such large-scale urban mining.

For much of the twentieth century, the industry depended heavily on two sources of lead that in industry parlance are labeled *primary* lead—lead sourced from virgin ore—and *secondary* lead—lead sourced from scrap. Both posed significant risks to the environment and public health. To understand how the lead industry closed the circle on lead-acid batteries, ensuring that almost all old batteries became new batteries by the 1990s while better protecting public health, requires taking a bird's-eye view of how lead was mined, processed, and, later, regulated.

Unlike the supply chains for most of the other batteries discussed in this book, the story of a lead-acid battery is centered in the United States. While lead has been mined in many places in the United States, Missouri has dominated US production from the eighteenth century to the present. As one local newspaper put it in 1901, Missouri's Lead Belt was "A Klondike at St. Louis' Very Door."[34]

Unlike other materials important to batteries, such as sulfur or lithium, which are mined in unconventional ways, lead mines are textbook underground mining operations. In Missouri's Lead Belts, miners excavated ore using room-and-pillar mining—taking the ore and leaving behind columns of rock to support the mine as they progressed. The rooms left behind could be cavernous, as illustrated in figure 1.2. By the 1920s, Missouri's Old Lead Belt was comprised of a few dozen major mining sites, many of which were linked together by an extensive warren of mining tunnels. As the story went, a miner could enter the mines near one mining town and exit in another town miles distant.[35]

Missouri's lead miners searched for galena, a silver-gray lead-sulfide ore. When they struck galena, not only could they see it, but they could feel its heft—it was three times denser than the surrounding rock. Surprisingly, galena was a relatively easy ore to mine. When galena broke, it cleaved at right angles, following the cubic structure that bound its lead and sulfur atoms together. Unlike other lead mining districts, where silver or zinc generated more revenue than lead, the Missouri Lead Belts were valued most for their lead content. Before the 1960s, Missouri lead mining was concentrated in the Old Lead Belt. After the 1960s, as the productivity of older mines declined, the industry transitioned operations sixty miles west to what became known as the New Lead Belt.

Today, the most visible clues to Missouri's long history of lead mining are old piles of waste rock and tailing ponds. Only a fraction of the ore hauled out of the Missouri's lead mines was galena; more than 95 percent was dolomite or other sedimentary rocks. The first step in processing lead ore was crushing it into progressively smaller pieces. In the Old Lead Belt, most ore was crushed to the size of small pebbles and separated by gravity.[36] In later years in the Old Lead Belt and entirely in the New Lead Belt, ore was ground into a coarse sand, mixed with water and chemicals, and separated using flotation tanks.[37]

Both processes left behind immense quantities of waste. In the Old Lead Belt, lunar-like mountains of waste rock, known as chat piles, loomed over mining towns such as Bonne Terre, Desloge, and Flat River. The biggest chat piles covered hundreds of acres and topped out at nearly two hundred feet in height. In the New Lead Belt, the slurry of sand leftover from flotation was pumped into shallow valleys. The most expansive tailing pond measured

FIGURE 1.2 Can you find the lead miner on the ladder? The underground room-and-pillar mines of Missouri's Old Lead Belt created a warren of underground excavations that extended for miles. This 1940 photograph is of St. Joseph Lead Company's Federal Mine No. 11. Southeast Missouri Lead District, Photograph Collection. 024301-1, The State Historical Society of Missouri.

more than one thousand acres in size. Many of these sites have since been listed as Superfund hazardous waste sites and slated for remediation.[38]

Crushing, jigging, and flotation all helped separate out lead ore, but those processes could not break galena's lead-sulfur bonds. That required heat. From 1882 until 2013, Saint Joseph Lead Company (known as Doe Run after a 1986 acquisition), which owned many of Missouri's lead mines, operated the nation's largest primary lead smelter on the banks of the Mississippi River, thirty miles south of St. Louis in the town of Herculaneum. The river provided an ample supply of water and made it easy to barge in needed supplies, such as coal and sand, and ship out refined lead. At its peak, the Herculaneum refinery produced more than two hundred thousand tons of lead per year.

Today we know that the lead industry imposed an enormous burden on the workers, their families, and the communities in which they operated. Although Saint Joseph's shipped lead by the ton to market, the lead, other heavy metals, and sulfur dioxide emissions left behind contaminated the community. Before the 1970s, however, few viewed Saint Joseph's operations with much concern. Instead, company towns such as Bonne Terre and Herculaneum praised Saint Joseph's for its corporate stewardship. As the historian Robert Faust explains, the company was not just "an employer but also an icon of generosity, responsibility, and stability."[39] *Fortune* magazine once labeled the company "Benevolent St. Joe."

While the Missouri Lead Belts and Saint Joseph Lead Company were mainstays of the primary lead industry, the growth of the battery industry created another important source of lead: scrap. Scrap dealers advertised for "battery lead plates" as early as the 1920s, promising to pay the "highest market prices at all times."[40] Unlike other major uses of lead, such as additives to paints or insecticides, the American Bureau of Metal Statistics observed that the use of lead in storage batteries was more of a "loan" than a form of "consumption." At the time, most car batteries lasted less than two years before needing replacement. That meant a steady supply of scrap lead.[41]

Unlike the primary lead industry, which extracted lead ore from rural mines in Missouri, Idaho, and other states, the secondary lead industry had a very different geography—it clustered in and around cities. Lead-acid battery recycling emerged as an early form of urban mining in the 1920s. The scrap industry was dominated by immigrants, particularly Jewish people from central and eastern Europe.[42] Hundreds of small-scale operators collected batteries and other scrap from service stations, garages, and scrap peddlers for resale to the lead industry. In Baltimore, a major center of metals reclamation, African Americans often did the most hazardous work—breaking apart batteries and smelting the metal.[43]

In retrospect, it is remarkable how little attention the lead recycling industry drew from public health officials in the early twentieth century. Although they were subject to smoke ordinances and early zoning regulations in cities such as Baltimore, the smelters were more often seen as nuisances than as urgent threats to public health.[44] Yet, these early smelters, which often had rudimentary furnaces and few pollution controls, were major sources of pollution, belching sulfur dioxide and heavy metals into the urban environment. Even today, it is possible to find elevated levels of lead, antimony, and other heavy metals in the soil near some former recycling sites.[45] Many of these former smelter sites have since been classified as Superfund hazardous waste sites. But before the 1970s, despite studies of the larger primary lead operations, few studies focused specifically on the smaller-scale recycling operations.

The hazards of reclaiming old batteries did not entirely escape public attention, however. A 1933 article in the *Journal of the American Medical Association* documented a cluster of lead poisoning cases among poor African Americans near scrap operations in East Baltimore. Public health officials had been alerted when a seven-year-old girl arrived at a Baltimore hospital unconscious, suffering from severe lead poisoning. She was the first of fifty-seven people diagnosed with lead poisoning, almost all of whom were Black.[46] Investigators discovered that the source of the lead was wooden casings from old batteries, leftover when the batteries were broken for recycling. Local junk dealers sold or provided the wooden casings free of charge to nearby families to use as stove fuel.[47]

During the Great Depression, with few options for keeping warm, African Americans living in center-city Baltimore burned what they could to cook and heat their homes. Without access to new electricity or natural gas networks that served other parts of Baltimore, they lived in what the historian Leif Fredrickson has described as an urban "energy desert."[48] The wooden casings from batteries were "considered to be excellent stove fuel."[49] But the casings were also a source of lead sulphate, which posed a potent health threat when the wood was burned, especially in rudimentary wood stoves. In response, Baltimore health officials prohibited the sale of battery casings and launched a public outreach campaign, which was replicated in other cities. The Baltimore lead-poisoning cluster was an early instance highlighting the dire threat recycling lead-acid batteries could pose to public health.[50]

But those concerns did little to slow the growth of the urban lead mining industry in the mid-twentieth century. Lead was easy to smelt, it required little capital investment, and it yielded a profitable commodity. No policies—either imposed by the government or organized internally by the industry—required the return or processing of lead-acid batteries prior to the 1970s. In fact, when the lead industry had tried to coordinate a network of battery recovery in the 1950s, to systematize the recovery of old batteries, it ran afoul of antitrust regulators in the Department of Justice.[51] That meant urban lead mining generally followed the price of lead, recovering more batteries when the price of lead was high.[52] In the mid-1960s, when the price of lead jumped by more than 30 percent in response to a strong economy and low production at mines, the battery recycling rate reached 97 percent. That suggested that under the right circumstances the lead-acid battery industry could close the loop on its supply chain.[53]

CLOSING THE LOOP ON LEAD-ACID BATTERIES

One might expect that the twinned concerns for lead pollution and environmental protection—both of which drew new attention in the 1970s—would drive a sharp uptick in lead recycling, transforming the industry into the closed-loop supply chain it prides itself on today. The story is not so simple. Three factors played a key role in transforming the lead-acid battery industry in the 1970s, pushing the secondary lead industry to the brink of collapse in the 1980s, before it reemerged as a highly regulated, consolidated, and large-scale recycling industry in the 1990s.

Factor 1: The New Maintenance-Free Battery
In 1968, a marketing firm tested four advertisements for the newly introduced "DieHard" batteries. They found the name generally resonated with consumers: it emphasized longevity and perseverance. As one respondent explained: "It will last and last and last. It may even outlive the car." Unsurprisingly, the study also revealed how little most consumers thought about car batteries. People were "in the dark" regarding the car battery. "I can't say too much about the battery in my car," explained another respondent. "Mostly, I take it for granted." What most distinguished batteries in the consumers' view was how long they were guaranteed for: four, five, or six years.[54]

That attitude toward batteries reflected a marked change from the 1920s and 1930s, when batteries were hardly taken for granted. At the time, the average car battery lasted less than two years before failing. Handbooks like Willard's emphasized the "care and attention" needed to ensure "long and faithful service."[55] By the mid-twentieth century, lead-acid batteries had become an unremarkable commodity distinguished largely by their brand names such as DieHard, Delco, or Exide. The invisibility of the lead-acid battery was, in many respects, a measure of the industry's success.

Drivers who looked under their hoods may have noticed a few changes in lead-acid batteries in the 1950s and 1960s. Plastic cases replaced the wooden or vulcanized rubber cases as plastics became cheaper and more readily available. Less apparent to consumers, but also important, plastic separators replaced the wooden separators inside the battery. Battery manufacturers also started marketing more 12-volt batteries, instead of 6-volt batteries. The change was driven by more complex electrical systems that included power windows, multispeed windshield wipers, cassette players, and support for higher-compression engines.[56]

The biggest improvement in the lead-acid battery industry came in the mid-1970s, with the advent of the maintenance-free battery. One of the weaknesses of almost all lead-acid batteries built prior to the 1970s was the need for regular additions of water to top off the electrolyte. Batteries still used an antimony-lead alloy for battery grids, which had been introduced in the 1920s and 1930s to make the grids easier to cast and more rigid. The downside was that the batteries were prone to self-discharge, and they gassed, meaning that they released small amounts of hydrogen when they were recharged.

Maintenance-free batteries offered significant benefits: they had a shelf-life that approached eighteen months; their cases were sealed, which eliminated fumes and prevented spills; and, most importantly, they did not need to be topped off with water. Although adding water to the electrolyte had become a routine step in servicing a car, it was a cumbersome step and contributed to dead batteries. Maintenance-free batteries were projected to account for more than 90 percent of the market by the early 1980s.[57]

The maintenance-free battery depended upon advances in the material chemistry and the manufacturing of lead-acid batteries. The advantages of calcium-based battery grids had been well-known since the 1930s: they were the preferred battery chemistry for large-scale applications in telephone switching stations, where low maintenance and reliability were highly valued. But before the 1970s, manufacturing lead-calcium battery grids was difficult and costly: the final concentration of calcium had to be just right—too much or too little compromised battery performance.[58]

In the 1970s, battery manufacturers solved this challenge by including trace amounts of other elements such as aluminum, silver, or bismuth, which prevented the loss of calcium, improved the workability of the lead-calcium alloy, and maintained battery performance. Those advances made it possible for battery manufacturers to scale up the production of calcium-based alloys and introduce "maintenance-free" batteries at a price that was competitive with traditional lead-antimony batteries. As one retailer advertised in 1979: "It never needs water. Ever."[59]

Although it is hard to know exactly what was happening inside battery factories, evidence suggests more automated manufacturing lines enabled the transition to maintenance-free batteries. Automatic grid-casting machinery increased the rate at which a worker could cast grids from forty grids per hour to five thousand grids per hour. Automated processes for pasting the grids with oxides also helped meet the tighter material specifications for the new batteries, while reducing worker exposure.[60] Yet, in the same years that the industry retooled for the maintenance-free battery, it also faced pressure to improve working conditions and address growing concerns about environmental pollution.

Factor 2: The Chronic Threat of Lead Pollution

For much of the twentieth century, those who lived and worked in lead industries voiced few concerns about the threat of lead poisoning. People seemed eager to believe that lead was the lifeblood of their economy, not a threat to their health and that of their family or community. A 1980 newspaper article about Missouri's Lead Belts noted, "Lead miners are amused by the common misconception that they are susceptible to lead poisoning."[61] And they were right: there was little immediate threat from mining galena. But when lead was milled, smelted, and recycled, or just left for decades in a waste pile, it became increasingly bioavailable, and with that came risks to humans and the environment. As scientific understandings of the scale and consequences of lead contamination began to change in the 1970s, much of the research focused on urban lead exposure. But those revelations had consequences that stretched out into the rural reaches of Missouri's Lead Belts.

Before the 1960s, the alarm over lead pollution focused largely on high-level, short-term lead exposure—that is, acute lead exposure. Starting in the 1960s, the threat of chronic lead exposure became a public crusade, as new scientific research demonstrated the ill consequences of long-term, low-level lead exposure. In *Lead Wars: The Politics of Science and the Fate of America's Children*, David Rosner and Gerald Markowitz explain how the issue galvanized public concern: "Environmentalists, politicians, community activists, conservationists, scientists, and public health officials understood that lead was one pollutant that challenged them all."[62] But making that case meant challenging the long-standing Kehoe-Kettering paradigm, which maintained that as long as lead exposure remained below a well-defined threshold, it posed no threat to human health.

Clair Patterson was an eccentric geochemist based at Caltech who specialized in the history of the Earth's crust. Patterson's research on the planet's age, research that had been decades in the making, depended foremost on his ability to precisely measure the concentrations of lead and other elements in geological samples. One problem stymied Patterson's research more than any other: every time he tried to date his samples, the concentrations of lead he measured were surprisingly high. Patterson came to realize his problem was contamination. Traces of lead measuring in the parts per million in his hair, on the walls of the laboratory, and inside the beakers used to prepare samples were enough to skew his analyses. Only by pioneering rigorous clean lab procedures and developing new research techniques to measure exceedingly low concentrations of lead did Patterson succeed in dating the age of the Earth to 4.5 billion years. That accomplishment drew him acclaim.[63]

Patterson was unsure where all that lead had come from, however. Robert Kehoe, the industry-supported lead toxicologist, had an answer: the low levels of ambient lead Patterson had observed were natural. Indeed, that was a basic tenet of the Kehoe-Kettering paradigm. But that explanation did not account for what Patterson discovered next.

In the mid-1950s, Patterson began to research the formation of Earth's crust in more detail. In researching the planet's geological history, he stumbled upon one of the most urgent pollution crises of the modern era. Patterson published a groundbreaking study in 1965 in the *Archives of Environmental Health* that reported that existing lead levels were not natural at all: they were the result of pollution. Lead pollution had become so ubiquitous that it appeared "natural" to an entire generation of scientists. He argued that lead contamination actually posed a "severe chronic lead insult" that threatened public health. It had been viewed with "complacency" for far too long.[64]

Patterson's research findings roiled the lead industry and its allies. Kehoe, who was asked to review Patterson's paper prior to publication, emerged as Patterson's leading critic. He saw in Patterson a young, brash crusader who had gone well beyond his expertise as a geologist to draw broad and unsubstantiated inferences about the threat lead posed to public health. As the industry began to reckon with Patterson's findings, Kehoe warned his colleagues, "The posture of the crusader is not the posture of the scientist."[65] The lead industry held private meetings to plot a response. They maintained that "all of the accepted medical evidence proves conclusively that the amount of lead particles to which the public is exposed—whether in the air they breathe or in the food and beverages they consume—is well within recognized limits of safety."[66]

Of course, much of that accepted medical research had been supported by the lead industry itself. That body of research had given little attention to the potential for low-level long-term lead exposure. Why would it? Such research was deemed unnecessary, as long as the ambient levels of lead present in the environment were considered to be "natural." Industry insiders believed Patterson's findings were so patently wrong that he should simply be dismissed as a "nut."[67] Kehoe and the lead industry's campaign of misinformation succeeded in casting doubt on Patterson's findings for more than a decade.

But Patterson's conclusions aligned with concerns other public health researchers had begun to raise about lead exposure in the early 1970s. At the time, the usual story of lead poisoning was that it was largely confined to the poor, immigrant, or African American communities in old cities where children lived in older, often dilapidated, housing where children had a tendency to eat lead paint. For this reason, the lead industry dismissed lead poisoning as an issue specific to "slum dwellings" and "relatively ignorant parents," not a matter of public health. In their view, until the nation could get rid of slums or educate the "ineducable parent, the problem will continue to plague us."[68]

Concern over childhood lead poisoning changed in two important ways in the 1970s. A 1975 study of a secondary lead smelter in Texas was one of several studies that made clear that lead dust posed as great a risk as did exposure to lead paint. When researchers took dust samples from homes neighboring the smelter, the closest homes had lead levels averaging 20,000 ppm (2 percent) lead. Even four miles away, lead levels in household dust samples averaged nearly 1,000 ppm. More troubling, the same pattern appeared in children's blood lead levels. Fifty-three percent of children living within a mile of the smelter had blood lead levels that exceeded 40 μg/dL falling to 18 percent of those living one to four miles from the smelter.[69] Children did not need to eat lead paint to be poisoned. Just living in a leaded environment put them at risk.

But what made that finding even more concerning was research that drew attention to the developmental consequences of long-term exposure to low levels of lead. A 1975 follow-up study on the Texas smelter revealed that children with high blood lead levels had significantly lower IQs and performed more poorly on reaction time tests, which suggested nerve damage.[70] A 1979 study further confirmed what many researchers had begun to suspect—that even children without clinical lead poisoning still suffered developmental and behavioral deficits. When researchers evaluated children's verbal abilities, reaction times, and classroom behavior, they found that, on every metric, children with higher lead exposure performed poorly compared to children with lower lead exposure, even if they remained below the threshold for lead poisoning.[71]

Such research prompted a decade-long effort to overturn the Kettering-Kehoe paradigm and impose regulations to protect the public, particularly children, from chronic, low-level lead exposure. The lead industry mounted a fierce campaign to discredit both the research that indicated low-level lead exposure posed a significant health risk and the researchers who advocated for tighter regulations on lead as a result. Although the lead industry succeeded in delaying some reforms or lessening their bite, between 1970 and 1990, scientists and regulators succeeded in enacting a series of restrictions on the use of lead in consumer products and new regulations on lead pollution from industrial processes that transformed the entire lead industry, including the lead-acid battery industry.[72]

Factor 3: A Cascade of Regulatory Action
Often occupational health regulations and environmental regulations are viewed as separate issues. For those on the front lines of the lead industry, those challenges were clearly intertwined. As the secretary-treasurer of the United Steelworkers Association explained in 1969, the "environmental consensus cannot stop at the plant gate. Our workers are demanding that you"— meaning Congress—"come inside with your legislative tools."[73]

In the working-class town of Herculaneum, Missouri, the Teamsters union urged regulations to "improve this most serious condition of smoke, gas, fumes, and dust which St. Joseph Lead Company is dumping onto the town of Herculaneum and surrounding areas."[74] In 1969, the "People of Herculaneum, Missouri" issued a joint letter signed by more than sixty residents complaining about the pollution.[75] Notably, these environmental activists were not wealthy suburbanites or outdoor enthusiasts; they were working-class residents living downwind of the nation's largest lead smelter. One resident put the conundrum plainly: "I realize this company creates jobs but it is destroying something much more valuable. I feel this company is violating the rights of every person who wants to breathe clean air."[76]

Such activism serves as an important reminder of the diverse concerns that propelled matters of occupational health and the environment into the national limelight in the late 1960s and early 1970s.[77] Starting in 1970, Congress passed a suite of laws creating the Occupational Safety and Health Administration (OSHA) and empowering the newly established Environmental Protection Agency (EPA) to protect workers, public health, and the environment.

This marked a key turning point for the entire lead-acid battery industry. Unlike the voluntary oversight that characterized the early industrial hygiene movement in the early twentieth century, led by Alice Hamilton and co-opted by Robert Kehoe, new laws like the Williams-Steiger Occupational Safety and Health Act (1970), Clean Air Act (1970), Clean Water Act (1972), and Resource Conservation and Recovery Act (1976) empowered government regulators to set health-based standards through command-and-control regulations, which industry had to comply with or face penalties.

Despite the efforts of industrial hygienists and the industry's voluntary efforts to protect workers earlier in the twentieth century, such changes were essential. In 1976, a battery manufacturing plant in Visalia, California, drew attention because, of the 150 people it employed, most of whom were Mexican Americans, 119 had sought medical attention related to lead exposure the previous year. As a union representative explained, "It's not right that we have to choose between our job and our health."[78] In Herculaneum, Missouri, Saint Joseph's had long escaped regulatory scrutiny on account of its reputation for corporate stewardship, but starting in the 1970s the smelter struggled to meet OSHA and EPA regulations.[79] And lead recyclers, especially small operations with few pollution controls, drew new concerns. One of many such companies was a small recycler in Dayton, Ohio, that recycled lead-acid batteries in a rudimentary furnace and stored the broken battery cases on-site in the 1980s. The largest pile of casings stood eight feet high and measured seventy-five feet in diameter. In some places, the soil was more than 5 percent lead where it had been soaked by battery acid and polluted from smelter fallout.[80]

The new laws required changes at every stage in the life cycle of lead-acid batteries. Between 1975 and 1986, the OSHA reduced the permissible level of lead in workplace air from 200 to 50 micrograms of lead per cubic meter of air. Starting in 1983, OSHA required monthly monitoring of blood lead levels and established a threshold of 50 micrograms of lead per 100 grams of blood for workers. By 1987, the EPA required smelters and battery manufacturers to implement the best-available control technologies under the Clean Water Act, and capped lead in waste water at 80 and 120 parts per million. By 1988, the EPA required smelters to keep average lead concentrations in the air below 1.5 μg/m³ at the smelter's fence line (to meet the ambient air lead standard adopted in 1978). Under the Resource Conservation and Recovery Act, spent lead-acid batteries were classified as hazardous waste, requiring new permits, groundwater monitoring systems, and liability insurance for facilities that stored lead-acid batteries.[81]

In short, the new regulations—driven by the new research on the consequences of lead—had implications for every step in the life cycle of a lead-acid battery, from mining to manufacturing to recycling.

A PERFECT STORM

In 1973, the United States initiated a phaseout of leaded gasoline. In 1978, it banned the use of lead paints. Children's blood lead levels began to decline, falling 37 percent between 1976 and 1980. As the use of lead in gasoline and paint collapsed, demand for lead fell for the first time in the twentieth century. That drove down the price of lead sharply. It fell from 41 cents to 19 cents per pound between 1980 and 1985.[82] Those low lead prices came just as the entire battery industry was faced with the task of retooling its operations, in part to manage the transition to maintenance-free batteries but largely to comply with the new occupational health and environmental standards, many of which required investing in pollution-control technologies.

This perfect storm led to an intense period of consolidation in the lead-acid battery and secondary lead industries. Consider the case of the lead recyclers. In 1975, eighty secondary smelters recycled lead. In 1979, secondary lead recovery approached seven hundred thousand tons, a new annual high.[83] By 1985, 40 percent of secondary lead capacity had shut down, and only twenty-three secondary smelters remained.[84]

The new occupational health and environmental regulations, instead of shoring up the lead-acid battery recycling industry, had left it "teetering on the brink of collapse," warned the industry.[85] In the mid-1980s, the recycling rate for lead-acid batteries dipped as low as 60 percent, which meant up to two hundred thousand metric tons of lead was going unrecycled each year.[86]

Lead-acid batteries by the ton were being improperly stored in warehouses, thrown in landfills, or, in some cases, illegally dumped or shipped abroad.

To the industry, such a piecemeal approach to environmental regulations exemplified the irrationality of government-driven, top-down environmental regulations. One executive fumed, "It is totally incomprehensible to me why the EPA would fight so hard over" emissions regulations "and then take the attitude that 200,000 tons per year of [battery] lead being improperly disposed of is insignificant and causing no harm."[87] The EPA's strategy did make sense: industrial lead pollution posed a much more immediate risk to human health than did unreclaimed batteries, because the trace pollution was more bioavailable. But, for the secondary lead industry, the recovery of batteries was as much a matter of economics as public health. Starting in the mid-1980s, they pressed for regulatory changes to stabilize the industry.

In 1987, the industry's trade group, Battery Council International (BCI), formally expanded its scope of activities to promote recycling. As its executive director explained, BCI had a "vitally important message" to convey—that "batteries are irreplaceable, that batteries are a social need, and that the manufacture of batteries and the recycling of batteries can be done, and is being done, safely."[88] Although BCI remained wary of more government regulations, in the late 1980s it did press for policies aimed at improving battery recycling, including a tax on primary lead, proposals that new batteries contain minimum percentages of recycled lead, and a ban on exports of lead-acid batteries.

Although such regulations to prop up lead-acid battery recycling never moved forward, states did move quickly to keep old lead-acid batteries out of landfills. In 1987, Rhode Island and Minnesota each passed laws promoting lead-acid battery recycling. BCI, fearing fifty different state policies, developed model state legislation in 1988. The key provisions of the BCI model law included a disposal ban on lead-acid batteries, standardized labeling on batteries and signage at retailers, and, most importantly, a take-back requirement at retailers and wholesalers.[89] By 1991, thirty-four states had passed such laws, most of which followed or adapted BCI's model legislation.

These state-level battery laws helped lay the groundwork for tolling arrangements, which were instrumental to closing the loop on lead-acid batteries. By banning battery disposal and instituting take-back policies, state laws helped shift the flow of spent batteries toward large retailers such as Sears Roebuck, Montgomery Ward, and K-Mart. In 1990, for instance, Exide and K-Mart struck a deal whereby Exide would pay two dollars for each battery returned to a local K-Mart. The companies described it as their effort to solve an environmental problem in a sensible way. Trucks transported new batteries from manufacturers to retailers, collected spent batteries along the route, returned the spent batteries to secondary lead smelters for recycling, where they picked up refined lead to take to the battery manufacturers, where the whole process began again.[90] As one scrap dealer described it in 1991, "It's now a closed loop for the battery."[91]

The new regulatory regime contributed to significant restructuring in the lead-acid battery recycling industry. By the 1990s, fifteen companies operating twenty-two secondary smelters remained. These smelters followed an integrated, "closed system" approach to secondary smelting that had first been pioneered in the mid-1970s. In 1976, National Lead explained its plan for what it described as its "closed system" for recycling lead: a fully automated, integrated secondary lead smelter capable of recycling whole batteries into raw materials for the battery industry. The primary feed for the plant would be spent lead-acid batteries, which the company described as a "natural resource" and the "basic raw material for this system."[92] The highly automated process minimized the risk for workers.

The system started by tilting entire trucks to an angle of 55 degrees, dumping batteries into an enclosed storage area. Front-end loaders then moved the batteries to a conveyor belt in bulk, which fed a crusher, which broke the batteries open to drain the acid. Next the batteries passed through an impact mill, which shredded the materials into small particles that could be sorted by flotation—the plastic case and separators floated to the surface, and metal fractions sank to the bottom. Most smelters then recovered the lead, sulfuric acid, and plastic for reuse.

Secondary smelters processed the metal fractions in a series of steps that included blast or reverberatory furnaces that applied heat directly and alloying or refining kettles that applied heat indirectly. Through the careful use of heat and chemical reagents, impurities could be removed, and a high-quality lead for grids or oxides produced.[93] Life-cycle analyses estimated that producing recycled lead required roughly one-quarter the energy as producing new, primary lead. That made secondary lead both environmentally and economically advantageous.[94]

Although the new smelters differed in their particulars, they increasingly followed the National Lead model: large-scale, integrated facilities that started with whole batteries and produced marketable, recycled commodities primarily destined for the lead-acid battery industry. Complying with environmental and occupational health regulations required significant investments in automation, including the battery-breaking systems, refining processes, containment buildings, and pollution-control technologies. National Lead described its integrated facility as an "assured disposal point for your scrap whole batteries and residues that will meet *all* of the environmental regulations our industry faces."[95]

By many measures, the restructured secondary lead system represented a marked improvement. In 1998, the nation's remaining twenty-two secondary lead smelters recovered one million metric tons of lead, nearly double what had been recovered in 1986.[96] Between EPA regulations, OSHA regulations, and liability issues, some argued that the regulatory scrutiny of the lead industry was second only to that of the nuclear industry. One industry executive described the regulatory environment as "relentless," but the industry believed that its successes in meeting these challenges put a "positive spin" on the industry.[97] Indeed, by many measures the industry had succeeded in closing the loop on the lead-acid battery. Since the 1990s, domestic recycling rates have exceeded 97 percent on average.[98]

One more consequence of the new environmental regulations bears mentioning: it also hastened a transition in the sulfuric acid industry. The invention of the hot-water Frasch mining process made the United States self-sufficient for sulfur starting in the 1910s. Since the 1930s, however, an increasingly important source of sulfur has been as a by-product, siphoned from the air exhaust of metal, petroleum, and natural gas refineries. After the 1970s, the Clean Air Act and subsequent amendments made the recovery of sulfur dioxide mandatory (sulfur dioxide is a noxious air pollutant and precursor to acid rain). As a result, domestic demand for sulfur was increasingly supplied as a by-product of air pollution–control technology or recycling operations, such as at secondary lead smelters.[99] The last Frasch mine in Louisiana closed in 2000, ending domestic production of mined sulfur.

The story of lead recycling points toward a conclusion that runs counter to the usual assumptions of environmentalists, who often favor small-scale, local operations. In the case of the lead industry, larger, highly consolidated, well-capitalized operations are preferable. The leakage of lead, whether in the form of batteries going to landfills or molecules of lead accumulating in the bodies of workers or neighboring residents, had all been significantly reduced. Such a tight secondary lead chain was not a product of industry, economics, or the peculiar qualities of lead alone: instead, it was a product of the interaction of these factors in a system driven by strict state and federal regulations.

THE LONG SHADOW OF LEAD

One of the signal accomplishments of public health advocates, environmentalists, and the lead industry has been the significant decline in lead contamination historically. That reduction contributed measurably to improving children's health nationwide. The most important driver behind this reduction was the elimination of lead in gasoline and paints, followed by reductions in industrial lead emissions. Between the 1970s and 2010s, average blood lead levels in the United States fell by almost 95 percent to less than 1 μg/dL. Although there is no safe level of blood lead, that level is well below 5 μg/dL, the Centers for Disease Control's current reference value for lead.[100]

Such declines are even more remarkable, considering the continued growth in the use of lead-acid batteries during the same years. By my estimates, the amount of battery lead circulating in the built environment—under the hoods of cars, in battery backup systems, and industrial batteries—grew from one million tons of lead in 1965 to nearly six million tons of lead in 2010, or from roughly ten to forty pounds of lead per person. With recycling rates approaching 100 percent, especially in developed countries such as the United States, it would be easy to chalk up the history of the lead-acid battery industry as a straightforward success story, with the decline of primary lead mining and the rise of recycling. But, as average lead levels began to decline nationally, it became easier to identify the hot spots of lead pollution that remained. That drew new attention to places such as the Old Lead Belt, the Herculaneum, Missouri, smelter, and lead recyclers, both shuttered and those still in operation.

The primary lead industry moved on from Missouri's Old Lead Belt in the 1970s, but the lead it left behind did not. In the 1990s, chat piles still framed the horizon. The mountains of mining waste slumped into people's backyards, served as sledding hills in wintertime, and the largest were popular sites for off-roading and other forms of recreation. Levels of lead in the chat and tailings averaged more than 3,000 parts per million, nearly three times higher than regulatory standards. Hot spots of lead contamination measuring more than 2,000 parts per million contaminated the local river and its floodplain as far as one hundred miles downstream.[101]

What had been the Old Lead Belt became one of the nation's largest Superfund hazardous waste sites in 1992, when the Big River Mine Tailings Site was added to the National Priorities List authorized by the Comprehensive Environmental Response, Compensation, and Liability Act (better known as the Superfund Act). In 1993, a representative of Fluor Corporation, which owned Doe Run at the time, described the risk posed by the waste sites as "nil to negligible."[102] Yet, 17 percent of children in the area had elevated blood lead levels, compared to 3 percent of children in a control community.[103] Soil surveys revealed that most schools and day care centers and many homes had lead levels exceeding permissible standards. Since the mid-1990s, regulatory action by the EPA required Doe Run, Saint Joseph's corporate successor, to remediate the chat piles and clean up residential yards, schools, and day care centers at a cost of over $100 million.[104] The work has been slow and litigious and is ongoing.

A century of lead production in the Old Lead Belt had polluted entire counties in rural Missouri. In contrast, the legacies of lead in the smelter town of Herculaneum were localized and acute. In Herculaneum, families and children lived and went to school within sight of the smelter's 350-foot stack. In 1978, under the Clean Air Act, the EPA set the limit on ambient level of lead in the air at 1.5 µg of lead per cubic meter averaged over three months— in its assessment, that was the maximum permissible level of lead that protected public health. In the fall of 1979, Saint Joseph's newly installed private fence-line monitor peaked at twenty times the EPA's newly set standard and averaged ten times the standard for the entire year. The company kept those findings to itself.[105]

Bringing the Herculaneum smelter into compliance with the Clean Air Act and other environmental laws proved a herculean task. By the early 2000s, after thirty years of regulatory action and $60 million in plant upgrades, the smelter barely met the ambient standard for lead pollution some of the time.[106] The Missouri chapter of the Sierra Club, which had begun to watchdog Doe Run's operations in the late 1990s, summed up the situation. "There have [been] 22 years of equivocation and delays in compliance by this company," they argued. "It is time to say enough: Clean up or get out."[107] In 2001, a public health study, initiated in response to shockingly high levels of lead contamination on Herculaneum's streets, revealed that 52 percent of children living within one-half mile of the smelter exceeded the standard for lead poisoning; 20 percent of children living within one to

1¼ miles did.[108] One state official commented, "I've never seen anything in the state this high or that even approaches this."[109]

Pressure on Doe Run to clean up Herculaneum mounted in the early 2000s. Two Herculaneum residents, Leslie and Jack Warden, who lived blocks from the smelter, served as lead plaintiffs in a 2004 lawsuit filed by the Interdisciplinary Environmental Clinic at Washington University in St. Louis that challenged the EPA for failing to update the national ambient air lead standards. The Wardens' son grew up in Herculaneum and suffered from lead exposure. The 1.5 µg of lead per cubic meter of air standard set in 1978 had last been reviewed in 1990, even though the agency was required to review standards every five years under the Clean Air Act. After a contentious court-ordered review process driven by new scientific research on the ill effects of low-level lead exposure, the EPA lowered the standard by a factor of ten, to .15 µg of lead per cubic meter in 2008.[110] The next year, Doe Run announced it would close the Herculaneum smelter by the end of 2013, ending more than a century of primary lead smelting in Herculaneum.[111]

The story was similar in the case of the secondary lead industry. A century of lead recycling had left behind a landscape of toxic hot spots, many of which were in or near cities. In 2001, William Eckel, a doctoral student at George Mason University, combined the tools of historical research and forensic geology to survey the legacies of the twentieth-century lead industry. His analysis of historic trade journals, city directories, and other sources revealed 660 sites likely associated with the lead smelting and recycling industries that were active at some point between 1931 and 1964 but had since been abandoned and, in some instances, reclaimed for other uses, such as gas stations, office parks, playgrounds, and schools. Geochemical sampling at many of these sites revealed the telltale signature of historical lead smelting: high soil concentrations of lead and antimony that posed a threat to public health.[112]

Not until 2012, when the national newspaper *USA Today* ran an extensive print and web-based exposé on what it labeled "ghost factories," did Eckel's work draw significant attention. Led by investigative reporters Alison Young and Peter Eisler, the newspaper followed up on the Eckel study, asking what steps the EPA had taken to investigate the 432 sites Eckel discovered that had been unknown to regulators. As Young explained, they discovered that in most cases governments had done very little. "Federal and state officials repeatedly failed to find out just how bad the problems were," the newspaper reported. "EPA and state regulators left thousands of families and children in harm's way, doing little to assess the danger around many of the more than four hundred potential lead smelter locations."[113]

The *USA Today* series led with the story of Tyroler Metals, a secondary lead smelter that operated in Cleveland, Ohio, from 1927 to 1957. In 2003, the site was investigated by the state in response to Eckel's initial research because of its proximity to a school and residential neighborhood. Investigators found soil lead levels ranging from 500 to nearly 2,000 ppm, which meant most samples were above the standard for remedial action.[114] Neither the EPA nor the state remediated the site or notified the school or nearby residents. One local resident, Ken Shefton, expressed his anger. "I've got a couple of kids that don't like to do nothing but roll around in the dirt," he told *USA Today*. "So, yeah, it is a real big concern."[115]

The publicity succeeded in spurring some action to address the legacies of historical lead smelters at the federal level. The EPA implemented a systematic review of the lead smelter sites between 2012 and 2017, identifying eighty-eight new sites for additional cleanup activity—responsibility for which was referred back to the states.[116]

The threats of the secondary lead industry were not confined to "ghost factories," however. The attention to the stricter 2008 limits for ambient lead under the Clean Air Act and the media coverage of Eckel's research on ghost factories helped draw scrutiny to secondary lead smelters still in operation. Exide operated one of two remaining smelters on the West Coast just east of Los Angeles in Vernon, California.

Exide had long avoided significant investments in upgrading pollution controls at its Vernon smelter. In 2013, regulators began to crack down on its operations, citing it for air emissions violations and improper hazardous waste management. "This should have happened a long time ago," said Roberto Cabrales, an organizer with a local environmental justice group, Communities for a Better Environment. "We're hoping that this will put Exide on the right path to either follow the rules or shut down if they can't."[117] In the 2010s, residents packed hearing halls and protested outside with signs that kept the spotlight on Exide's record of noncompliance: "Stop Exide: Serial Polluter!" "Exide: Lead Kills Kids!"[118]

The regulatory scrutiny and pressure from local environmental justice activists and community members resulted in a cascade of temporary closures and promised upgrades. In October 2013, despite having filed for bankruptcy, Exide agreed to upgrade its emissions-control technology, commission dust and soil testing in nearby neighborhoods, pay for blood lead tests for local residents, and comply with more rigorous monitoring.[119] The soil testing revealed extensive contamination: only 2 percent of yards had lead levels that posed no risk; and nearly one-third had levels that posed a risk to children's health.[120]

In August 2014, the Department of Justice opened up a federal criminal investigation based on Exide's history of environmental violations. That forced Exide's hand. Exide agreed to demolish the plant, remediate the site, fund the cleanup of homes in the vicinity of the plant, and pay a $50 million fine to avoid prosecution.[121] Local residents and activists believed those actions were long overdue. They were right: over the previous twenty years, regulators had documented frequent and recurring violations of regulatory standards—hazardous waste spills, faulty containment systems, high levels of lead contamination—many of which required immediate redress. Yet, state regulators had fined the company only seven times and had allowed it to continue operating.

Taken together, these examples point to a pattern of regulatory failure in response to the threat of lead pollution, both historically and more recently. Such failures drew national headlines in 2015, when local residents and outside researchers discovered that drinking water in Flint, Michigan, had been contaminated with lead. Although local, state, and federal EPA officials were aware of the threat to public health in Flint, they proved slow to act, with the EPA failing to intervene for six months after it had the necessary information.

Most alarming, as in Flint, the communities on the front lines of the lead industry were and often remain disproportionately people of color and the poor. In Cleveland, the neighborhood closest to Tyroler Metals was largely African American. In Los Angeles, the neighborhoods near Exide were almost entirely Hispanic with above-average rates of poverty, low home-ownership rates, and high proportions of non-U.S. citizens. And those clustered around Missouri's Lead Belts were disproportionately white and working class. Although all Americans may count on lead-acid batteries to start their cars, history makes clear that it has been communities of color and poorer working-class Americans who have disproportionately borne the burdens of the industry.

The unjust relationship between race, class, and the environmental hazards of a "dirty" industry, such as the lead industry, while alarming, would not come as a surprise to anyone who has studied environmental justice. But these concerns are just as urgent in the case of seemingly "clean" energy technologies too. The systems of extraction and manufacture important to solar panels, wind turbines, and lithium-ion batteries will remake geographies of extraction and production, and increasingly those burdens are borne by disadvantaged communities outside of the United States.

IMPORTING METALS AND EXPORTING HAZARDS

Throughout the twentieth century, the United States largely met demand for lead-acid batteries domestically. The United States led the world in lead mining, lead-acid battery manufacturing, and lead recycling. American corporations such as Exide, East Penn, and Trojan dominated the industry. In part, this was because the United States had large domestic reserves of lead, centered in Missouri. In part, this was because lead was heavy, which gave an advantage to domestic manufacturers. But as occupational health and environmental regulations increased the costs of doing business and the costs of global shipping fell, the geography of the lead-acid battery industry changed.[122]

The closure of the Herculaneum smelter marked a historic turning point for the lead industry. When it closed, it was the last remaining primary lead smelter in the United States. Although major lead mines continued operations in Missouri and Alaska, the ore was shipped abroad for smelting. That means since 2013, all primary lead has been imported.[123] One country that helped meet some of that demand was Peru, which saw its exports to the United States more than quadruple in the early twenty-first century. As troubling as the legacies of lead in Herculaneum, Cleveland, or Los Angeles were, events at Doe Run's La Oroya, Peru, smelter between 1999 and 2009 were even more devastating.[124]

La Oroya, Peru, had a long history of metal mining and smelting, starting in 1922, when an American firm set up operations in the region. In the case of La Oroya, more than half a century of unregulated smelting of copper, zinc, silver, and lead had left the small, mountainous town laden with the fallout of heavy metals and the surrounding hillsides scorched by acid deposition. In the 1990s, the La Oroya smelter complex released 2.5 tons of lead and nearly one thousand tons of sulfur dioxide per day from its main stack— pollution levels that were roughly ten and twenty times higher than those in Herculaneum in the mid-1990s.[125] A 1999 toxicology study revealed that the average blood lead level of 343 children tested in La Oroya was 33.6 µg/dL, with more than 86 percent presenting values exceeding 20 µg/dL. Children living closest to the smelter had an average blood lead level of 43.5 µg/dL. That put all of La Oroya's eight thousand children at high risk.[126]

FIGURE 1.3 The smelter in La Oroya, Peru, looming in the background, posed a significant health threat to the community, especially children. Courtesy of Keith Dannemiller.

In the early 1990s, Peru undertook a privatization program as part of a broader turn toward neoliberalism, undoing the socialist legacy of its 1970s military government, that included selling off state-owned oil, energy, telecommunications, and mining companies. Just before Peru auctioned off the La Oroya smelter complex, the US-based magazine *Newsweek* warned, "If anyone's interested, hell is up for sale."[127] The eventual buyer was Doe Run Peru, which was owned and run by the same New York–based company that owned the Herculaneum smelter in Missouri. It purchased the smelter for $247 million and committed to a $120 million program plan to modernize the plant and reduce pollution.[128]

Although Doe Run Peru would be vilified for its efforts, the company made substantial investments in La Oroya in the early 2000s, including plant upgrades, worker hygiene programs, and a community outreach initiative. It launched programs to lower blood lead levels by promoting what it described as "a culture of hygiene and health intended to decrease blood lead levels" especially among children.[129] Ambient airborne lead pollution fell by more than 50 percent, and, according to Doe Run's analyses, average blood lead levels in children dropped to 26.1 µg/dL.[130] Despite Doe Run Peru's improvements, however, it did not undertake the more expensive retrofits that promised the greatest reductions in pollution.

Public pressure mounted for action to clean up La Oroya. In 2006, the Blacksmith Institute placed La Oroya on its inaugural list of the ten most-polluted places in the world.[131] And in the winter of 2006, *Mother Jones* ran a feature on Doe Run, drawing parallels between the company's operations in Missouri and those in Peru.[132] This surge of international attention to conditions in La Oroya stemmed from the work of US-based Protestant missionaries, who responded to a plea for help from local evangelicals in La Oroya in 2002. Their partnership led to what one US-based missionary described as a "transnational campaign for children's health in La Oroya."[133]

Doe Run never followed through fully on its commitment to upgrade operations in La Oroya. In the wake of the 2008 financial crisis, Doe Run Peru filed for bankruptcy. Peru reclaimed the smelter as Doe Run Peru's largest creditor. By the end of 2017, 1,600 citizens of La Oroya, primarily children, were party to lawsuits filed in state and federal courts in Missouri alleging damages from Doe Run's Peruvian operations.[134] Despite Doe Run's extensive efforts to dodge liability for pollution in La Oroya, those suits continued to proceed through US courts in 2020. In turn, Doe Run sought legal redress for what it deemed to be Peru's illegal seizure of its operations.[135]

The crisis in La Oroya reflected a broader trend in the lead industry. Less US primary lead production meant more imported lead from countries like Peru, South Korea, and China, all of which had weaker standards for protecting workers and public health. Between 2000 and 2010, lead smelting in China quadrupled from one to four million tons per year, with primary lead smelting accounting for nearly two-thirds of production. During that time, China became the largest destination of lead ore concentrates exported from the United States. Local protests in China, scholarly research, and international media coverage drew attention to problems with lead pollution in China too.[136]

Closing secondary lead smelters in the United States in the early 2000s also pushed more lead-acid battery recycling abroad. This had first been a concern in the 1980s, when the US lead recycling industry nearly collapsed. Scrap batteries from the United States began turning up in countries such as India, Mexico, South Korea, Venezuela, and Taiwan. In 1990, PBS's *Frontline* aired an episode titled "The Global Dumping Ground" that focused on the international trade in hazardous waste. It spotlighted the ill consequences of recycling US batteries in Taiwan.[137] Although restrictions were adopted on international trade in hazardous waste in the 1990s, including the Basel Convention on the Control of Transboundary Movements of Hazardous Wastes and Their Disposal, those regulations initially included a loophole for materials destined for recycling, such as lead-acid batteries.[138]

The largest foreign destination for spent US batteries has been Mexico. Exports surged from 4,300 metric tons of lead in 2000 to 200,000 metric tons of lead in 2010. Secondary smelter shutdowns in places like Pennsylvania, Texas, and the Exide smelter in California likely contributed to the growing transboundary trade. Johnson Controls' acquisition of two Mexican smelters in the early twenty-first century, contributed to the surge of battery exports. Johnson Controls, the largest domestic lead-acid battery manufacturer at the time, described their acquisitions as part of a strategy to shore up their access to recycled lead.[139] Many observers, however, including those in the lead industry, environmental health advocates, and economists, viewed the shift to Mexico as a race to the bottom driven by trade liberalization.[140]

Under the terms of the North American Free Trade Agreement, member nations were encouraged to meet the same science-based occupational health and environmental standards. But that requirement was unevenly enforced.[141] Mexican standards for lead were lower and less thorough than those in the United States: occupational exposure limits were three times the US standard (150 $\mu g/m^3$); ambient pollution levels were 1.5 $\mu g/m^3$ (the same as the US standard prior to 2008); and worker blood lead levels were unregulated.[142] The *New York Times* reported in 2011 that the "rising flow of batteries is a result of strict new Environmental Protection Agency standards on lead pollution," which made it more expensive to recycle in the United States while doing nothing to prevent "exporting the work and the danger to countries where standards are low and enforcement is lax."[143]

Public health activists led by US-based OK International and Mexico-based Fronteras Comunes succeeded in drawing substantial attention to the issue of lead-acid battery exports in the early 2000s. Their goal was a ban on exports. The NAFTA-created Commission for Environmental Cooperation undertook an in-depth investigation, faulting the weak regulatory standards in Mexico.[144] US-based secondary lead smelters voiced objections to the exports: "Every day that scrap batteries are exported from the U.S. to Mexico and other nations is another day that the U.S. poisons our neighbors' children."[145] Public health economists documented a relationship between the increase in lead exports to Mexico and poor health outcomes for Mexican children living in proximity to the smelters.[146] Yet, the public pressure succeeded only in improving regulations for documenting the trade in spent lead-acid batteries, not imposing limits on such exports.[147]

The most troubling fate of lead-acid batteries is the one that is hardest to follow: the flow of batteries to countries least well-equipped to recycle them safely. Compared to the volume of batteries recycled domestically or the volume of batteries exported to Mexico, the percentage of US batteries that go to recycling operations in other countries is relatively small. Yet, around the world, informal recycling operations, where lead batteries are smelted in open barrels or rudimentary furnaces, remain common. The small scale of such operations, their wide distribution, and the rapidity with which they are erected and dismantled means there is no comprehensive information about the informal lead-processing industry.

Such sites often draw attention only in moments of crisis. For instance, in the 1990s, a lead smelter near the port of Haina in the Dominican Republic resulted in an epidemic of local lead poisoning, with more than 90 percent of nearby residents testing above accepted levels. The site was described as the Dominican Chernobyl.[148] In 2007, World Health Organization authorities became aware of an epidemic of lead poisoning outside Dakar, Senegal. Locals had been collecting lead fragments leftover in the sandy soil from battery recycling operations. Children often played in the sand. Sometimes adults brought soil to their homes to sieve and bag. Such activities resulted in lead exposure connected to the deaths of eighteen children in 2007 and 2008.[149] In 2015, Phyllis Omido, a Kenyan, was awarded the Goldman Prize for environmental activism in recognition of her efforts to close a secondary lead smelter that operated with few precautions and sickened her son and her community. She was jailed for her efforts and violently attacked by two armed men. Ultimately, she was successful, forcing the closure of the smelter in March 2014.[150]

When the consequences of such small-scale and unregulated lead recycling operations are factored in, the global consequences of recycling lead-acid batteries are astounding. Jack Caravanos, a professor of public health and lead researcher at the Blacksmith Institute, warned that "lead poisoning from improper automotive battery recycling activities is the number one childhood environmental health threat globally."[151] In few cases does the exposure result in death. Instead, lead's toll is measured in diminished mental capacity and chronic health ailments that last a lifetime. It is estimated that one in three children worldwide suffer from elevated blood lead levels. Researchers surveying the literature on exposure from sites of lead pollution identified metal foundries and battery recycling as "more than likely some of the major contributors to hotspot lead poisoning worldwide."[152] And as the use of lead-acid batteries in many countries continues to grow, to service electric bikes, for small-scale power backup for solar arrays, and for automobiles, the challenges of safely managing lead-acid batteries at end of life are likely only to grow.

It would be fitting to end this chapter with the death of the lead-acid battery. California and the European Union have broached banning the use of lead in batteries in recent years, on account of toxicity, the persistent threats to public health, and the availability of alternatives. Just as lead was phased out of gasoline, paint, and water pipes, activists argue that the time has come to phase lead out of batteries too. Even if governments don't act, some expect shifts in the market to do the job. In 2019, *Forbes* magazine pronounced that "Lead-Acid Batteries Are on a Path to Extinction," soon to be overtaken by lithium-ion batteries and other newer, safer, more efficient technologies.[153]

Despite the persistent threat of lead exposure, we shouldn't lose sight of the big picture. Compared to not recycling at all, the story of recycling lead-acid batteries has been a success. Between 1960 and 2010, I estimate that the United States relied on forty million metric tons of lead to power roughly five billion lead-acid batteries, most of which were the starting and lighting batteries that made cars, trucks, and other vehicles easy to start and convenient to run. Even when factoring in lower recycling rates historically, recycled lead still accounted for two-thirds of domestic lead consumption.[154] That is far preferable to the alternative. If the United States had not recycled lead, primary lead production would have been three times greater, likely required far more imported lead, and resulted in significantly higher energy consumption, pollution, and toxic waste.

But the history of the lead-acid battery makes clear the price that communities on the front lines of the lead industry have paid as part of this "environmental success story."[155] As lead-acid battery recovery nears 100 percent in countries such as the United States, the most advanced secondary lead smelters operate with pollution control systems that exceed regulatory standards, and blood lead levels for workers have fallen sharply, the industry seems to offer a model for a circular economy, in which spent products become the raw material for new products, offsetting resource extraction, energy consumption, and avoiding pollution. Yet, when processing billions of pounds of lead each year, even when recycling rates are high and pollution controls efficient, the small fractions of batteries that go unrecycled or chemicals released to the environment still pose grave risks, especially for poor communities and communities of color living near smelters. In the United States, secondary lead recyclers still reported releasing 1,200 pounds of lead into the air and disposing of nearly one million pounds of leaded waste to on-site or off-site facilities on average in 2019.[156] And the performance of domestic recyclers far exceeds those in many other parts of the world, especially the little-regulated informal recycling industry.

Continuing to scrutinize the lead-acid battery recycling industry is essential, for it is unlikely that lead-acid batteries will relegated to the dustbin of history. The most obvious reason is that demand for lead-acid batteries is likely to continue to grow as demand for conventional cars, electric bikes, and small-scale battery backup systems continues to expand, especially in developing countries. But some researchers anticipate the market for lead-acid batteries could even accelerate, driven by a new generation of lead-acid batteries that employ newer battery management systems and advanced electrode materials to perform at levels on par with lithium-ion batteries for stationary applications, such as grid-scale electricity storage. In a world where battery demand is set to grow exponentially to support a clean energy future, and concerns about resource availability are mounting, a new generation of lead-acid batteries—with their material abundance, low cost, and high recyclability—could play an essential role in the "future portfolio of energy storage technologies."[157]

Consider the alternative: what if we did phase out lead-acid batteries entirely? Despite the visions of a fossil fuel–free future, powered by renewable energy and supported by lithium-ion batteries, we cannot overlook the legacies of the past we hope to leave behind. This is a second reason to be cautious about banning lead-acid batteries. What would happen to the millions of tons of lead already circulating in the built environment globally? Would we landfill it? Return it to old lead mines? What if bans go into effect only in some countries? That could concentrate the life cycle of lead in other, likely less-developed countries, exacerbating the already existing inequities in the distribution of lead hazards.

The real challenge is not banning the use of lead in batteries but, rather, finding ways to ensure that the lead already in circulation worldwide is safely managed to best protect both human health and the environment in perpetuity. New hydrometallurgical lead-acid battery recycling processes that replace high-temperature smelting with water-based chemical processing could be far less hazardous than traditional lead recycling. And long-lived applications, including a new generation of advanced lead-acid batteries and, potentially, a new high-performance perovskite solar cell (of which lead is a small, but fundamental, component) offer long-lived applications for lead.[158] Despite lead's hazards, charting a course toward a clean energy future may mean giving lead a new lease on life, and finding ways to safely use the millions of tons of lead we have already liberated and can never entirely leave behind.

2

AA BATTERIES AND
A THROWAWAY CULTURE

THIS CHAPTER BEGINS WITH A CONFESSION: I THROW MY OLD SINGLE-use batteries into the trash. It was not always this way. I used to dutifully collect my spent AAs, AAAs, and 9-volt batteries in an old yogurt tub, which I would then take to the nearest recycling center, usually in town or at the college where I teach. But a few years ago, instead of dropping off my spent batteries for recycling and going about my day, I tracked my batteries to their eventual destination. It was a long trip for the batteries.

Here is what I learned: My spent batteries were shipped from Massachusetts to California and then to a high-volume recycling operation in Canada. There, the batteries were sorted, ground into a powder, and treated mechanically to recover steel for new metals and low-grade zinc, manganese, and potassium, which were used as micronutrients in agricultural fertilizers. The paper and plastic were burned to generate energy that helped power the process. The recycling process was efficient: the recycler claimed "100 percent" capture. The problem was the shipping. Transporting spent batteries 5,500 miles to recover small amounts of metal and fertilizer did not make much environmental sense. In fact, recycling single-use batteries often uses more energy than it saves, and the material and health benefits of doing so are modest.[1]

Most troubling, focusing on where my batteries went at end of life distracted me from an even more important question: where did the batteries come from in the first place? It turns out that the vast majority of the impacts of a single-use battery are concentrated in sourcing and processing the raw materials needed to make them. What is required to manufacture a single-use battery is shocking. It takes roughly 160 times more energy to manufacture a typical single-use battery than the battery returns during use. That makes the carbon footprint of the average single-use battery more than forty times greater than an equivalent amount of electricity generated by a coal-fired power plant. Sourcing the metals for a battery that weighs less than an ounce can require mining a pound or more of ore.[2] And all of this high-quality processing is costly. The supply of electricity delivered by a single-use battery costs more than one thousand times the equivalent amount of electricity delivered by the electrical grid.[3]

Yet, over the course of the twentieth century, Americans purchased, used, and disposed of single-use batteries by the tens of billions. Their portability, storability, and ubiquity made them a key enabling technology in the rise of twentieth-century consumer culture, powering generations of portable flashlights, toys, cameras, remote controls, and other consumer electronics. While the world may hope that we can close the loop on lithium-ion batteries, as has increasingly been the case with lead-acid batteries, the challenges of manufacturing and recycling single-use batteries are instructive. This chapter is organized around three questions that are helpful in thinking about prospects for a battery-powered future.

First, what can the history of the single-use battery teach us about the importance of standardization? Today, each electric car manufacturer depends on its own proprietary battery systems and specialized battery chemistries. The 1910s and 1920s were the "Wild West" for single-use battery manufacturers too. By the 1940s, however, government and industry had partnered together to establish standards for the size and performance of single-use batteries. By the end of the century, single-use batteries had become so commonplace, predictable, and interchangeable that they amounted to a form of consumer infrastructure, undergirding a wide range of modern consumer applications.

Second, how can the history of the single-use battery help us think about the complex and interconnected history of modern materials? Unlike the lead-acid battery, which is the largest end use of lead, single-use batteries depend upon a wider array of metals and chemicals, all of which are used in much larger quantities by other industries. That means the story of disposable batteries does not hinge on any single metal or chemical. In fact, over time, its material composition has become increasingly complex, even as it depends upon ever-more-pure materials. These changes contributed to two key transitions in the history of single-use batteries: the transition from zinc-carbon to alkaline-manganese batteries and the elimination of mercury from disposable batteries. That latter transition, which came about in the 1980s and early 1990s, sharply reduced the amount of mercury in the nation's waste stream, making it a little-known environmental success story.

Third, why does recycling still occupy an outsized place in discussions over single-use battery policies? Although single-use batteries are nontoxic—largely as a result of phasing out mercury—and the environmental costs of recycling single-use batteries largely outweigh the benefits, recycling remains a focal point of consumer activism and policy discussions. In 2014, Vermont mandated that manufacturers take back used single-use batteries. In California, it is illegal to throw them in the trash. And in 2015, Energizer began advertising new batteries containing 4 percent recycled content. Yet, the evidence suggests that the biggest step toward advancing sustainability is not improving the management of single-use batteries at end of life but, rather, curtailing their production in the first place.

To address these questions, this chapter traces the environmental history of the single-use battery from its invention in the early twentieth century to efforts to recycle them at the start of the twenty-first century. While disposable batteries have always delivered a vanishingly small amount of electricity, they have played a pivotal, but little-appreciated, role in enabling a modern culture of mobility. From powering farm radios to providing the spark for early automobiles to powering flashlights, this portable source of power has enabled people to transcend both time and space—making possible instantaneous communication, ensuring reliable transportation, and illuminating the darkness. The significance of disposable batteries has never turned on the amount of energy they supply, which is vanishingly small, or their energy efficiency, which is exceedingly low. Instead, it is the quality of the energy—its instantaneity, portability, and storability—supplied by AA, AAA, and 9-volt batteries that make them so useful. This chapter makes clear the costs of that convenience.

A SHORT HISTORY OF THE MANY USES
OF DISPOSABLE BATTERIES

The market for dry-cell batteries grew quickly at the start of the twentieth century. Before the electrical grid reached most people, disposable batteries were, as the editors of the *General Electric Review* reported, "the commonest source of electricity."[4] What enabled the rapid growth of the dry-cell market is signaled by its name. Unlike earlier batteries, such as the Leclanché cell, which contained a spill-prone liquid electrolyte, dry-cell batteries contained a paste of manganese dioxide, ground carbon, and sawdust or other material saturated with the electrolyte all sealed in a zinc canister. That made the dry-cell battery easier to transport and provided more reliable service.

A cascade of inventions and competing technologies contributed to the rise of the telegraph and then the telephone in the nineteenth century. Even as these seemingly magical technologies collapsed time and space, they shared one thing in common: they all relied on batteries. The power supply for early telephones was local—meaning it was located in the telephone, not supplied over telephone lines. Historically, a telephone receiver rested on a hook. Removing the receiver from the hook closed a battery-powered electrical circuit. Speaking into the telephone modulated the circuit's current, which was carried over the wires. On the receiving end, the faint variations in current caused a speaker to vibrate, "exactly reproducing the vibratory motion of the transmitting diaphragm." In that way, early telephones depended upon the action of the batteries.[5]

A 1905 manual for telephone systems included an entire chapter dedicated to primary batteries. It made clear how fragile and temperamental early batteries were before the advent of dry cells. It detailed the steps necessary to assemble, maintain, and troubleshoot early battery systems. That means that twenty-first-century smartphone users have something in common with the very earliest of telephone users—each fretted about their batteries. Early telephone subscriptions included regular battery service. "A minute's work" to maintain a battery, explained one observer, "saves more trouble in the future and complaints from subscribers about poor service."[6]

Dry-cell batteries did not require maintenance; they just required replacement. That reduced maintenance costs for telephone companies and provided a steady business for battery manufacturers. As demand for telephones grew, so too did demand for the batteries to power them. Philip Nungesser, who founded a company that would eventually become a part of Energizer, acknowledged "the courtesy of YOUR industry"—meaning the telephone industry—"in helping ME to put MY business on the map."[7] In 1913, the #6 Nungesser Dry Cell Battery for telephone service cost about six cents to manufacture and sold for twenty-five cents (about seven dollars today). The company promised, "With Nungesser Dry Cell Batteries your telephone service will be A-1 at all times."[8] Dry-cell battery sales were estimated at seventy-one million per year, industry-wide, by 1914.[9]

Early automobiles also relied on dry-cell batteries. Before the electric starter motor became common in the 1910s, disposable dry-cell batteries helped to start and, in some cases, run early gasoline cars. Most early Ford vehicles depended upon dry-cell batteries for starting and ignition. A set of batteries could keep the car running for 100 to 200 miles before they had to be replaced. When the Model T went into production, dry-cell batteries helped start the car. The driver knew the battery ignition system was working because of the hum of electricity they could hear in the ignition coils. Once started, the driver could switch over to a magneto-ignition system that was powered by the vehicle's engine. In short, all of this meant that early automobiles required more than just gasoline to run; they required a fresh set of batteries.[10] The French Battery Company promised its heavy-duty Ray-O-Spark battery would deliver "thousands of fat, crackling sparks always ready to pop—white hot—into a fuel charge, exploding its powerful driving energy, hours on end day after day."[11]

Batteries played a pivotal role in one other early twentieth-century technology: the rise of the radio. The first commercial radio broadcasts began in 1920. A decade later, twelve million homes in America had radios, far outpacing the adoption of the telephone or electrification of homes. As an executive at the Radio Corporation of America put it optimistically, radio is leading a "march toward the easier exchange of ideas."[12] Radios relayed news, weather, and entertainment. In the 1930s, President Franklin D. Roosevelt delivered his weekly "fireside chats" to the nation via radio broadcast. New entertainment and variety shows seeded the beginnings of a national mainstream culture. As one farmwife put it in 1925, "The radio service is a godsend to isolated families."[13]

Early radios depended on three different batteries to power different parts of the radio at different voltages. In some cases, radios relied on a lead-acid storage battery for their power. But they were expensive, messy, and required recharging (which was not easy without reliable electrical service). Because dry-cell batteries were relatively inexpensive, easy to connect, and easy to

For Fat, Hot Sparks

Outside, batteries have much the same appearance. But inside —*what a tremendous difference there is!* That's the reason for the exceptional performance records of French Dry Batteries.

To give fat, hot sparks is the only ambition these batteries have. Everything that goes into them, from the zinc casings to the carbon pencils, is scientifically selected for this one specific purpose—*to give fat, hot sparks for the longest time.*

Consequently, French Dry Batteries—Package Electricity—

give ignition service vastly superior to the ordinary.

The "shelf life" of French Dry Batteries is unusually long. That is, they will stand unused for months with practically no deterioration. Their power does not "leak" away. They stay fresh, full-powered.

For ignition work especially, and for all other dry battery uses, it is a decided advantage to install French Dry Batteries. They are acknowledged better batteries. Your dealer has them —in the familiar blue carton.

FIGURE 2.1 Early gasoline-powered vehicles relied on "fat, hot sparks" from disposable batteries to run the engine. *Saturday Evening Post*, March 15, 1920.

ship, they were "a marked favorite," explained the 1921 book *Radio for Everybody*.[14] All battery manufacturers promised that their batteries had the longest life, were least prone to leak, and delivered the best reception. They urged consumers to pick out "a battery that is free from the hissing, sputtering, and frying noises that are so often confused with static."[15] For the first time, manufacturers began advertising their products with "batteries included."

What is remarkable is how quickly advances in technology overtook each of these key uses of batteries: centrally powered phone systems replaced local telephones; the lead-acid starter battery and motor replaced the dry-cell ignition system in cars; and the advent of the electrical grid made plug-in radios possible. The largest and longest-lasting market for dry-cells emerged with the invention of the flashlight. Portable sources of light were hard to come by before the early twentieth century. Most people relied on candles or lanterns. In comparison, a flashlight "does not flicker in a draught, extinguish in the wind, and is controlled instantly by finger pressure," advertised American Ever Ready Works. It "is the light that everyone needs." Despite those advantages, flashlights met with limited success at first. Weak bulbs and short lives meant they were "little more than ingenious but expensive toys."[16]

Two advances made flashlights practical by the 1910s. First, the invention of a more efficient tungsten filament (what made a flashlight bulb glow) in 1907 yielded three to four times more light per watt. That meant a battery-powered bulb could cast a reasonably useful light. Second, special flashlight batteries extended the service life of flashlights by 500 percent.[17] The American Ever Ready Company, now known as Energizer, advertised *101 Uses for an Eveready* in 1917. They ranged from providing a reading light to medical diagnoses to frog spearing.[18] In 1922, flashlight sales were five million per year, and flashlight battery sales were five times that.[19]

The early industry faced three problems, however. Early batteries had a shelf life of weeks or months, making them a perishable product notorious for going bad. Poor battery seals meant that batteries often leaked electrolyte (despite manufacturers' claims), which could damage the radios, flashlights, and other devices they powered. And, finally, every new invention seemed to require a new battery size, which created confusion for manufacturers and consumers alike.

STANDARDS: FROM THE NO. 6 TO THE AA BATTERY

In 1909, the *Electrochemical and Metallurgical Industry* surveyed the nascent dry-cell battery industry. It saw disorder. Annual output had reached forty million cells—primarily to supply telephones, telegraphs, and vehicles—but it warned that the "general lack of knowledge about the dry cell is appalling, while the usual methods of rating dry cells remind one at best of kindergarten methods for making electricity easy."[20] It reported that the American Electrochemical Society had set about the task of establishing standards for testing battery performance. Without such tests, they worried that consumers had no assurance of battery performance. Despite the fact that most manufacturers relied on the same battery chemistry, the resulting batteries "may be good or they may be practically worthless."[21] In the engineers' view, the industry had "progressed to such a point that a reliable product is obtainable."[22]

The US Bureau of Standards issued the first official specifications for dry-cell batteries in 1919, in the wake of World War I. The government specified that dry-cell batteries be "easily portable," be "nonspillable" in any position, emit "no gases or other products," and be immediately usable once installed. The Bureau set forth performance tests for common applications, including ignition, telephone service, and flashlights. It also issued tentative specifications for the physical dimensions of batteries, codifying existing practices. "By far the most common" in the early twentieth century was the No. 6 battery. It was six inches high, two and a half inches wide, and weighed two pounds. These large-format batteries were used in telephones and ignition systems. The 1919 standards also described a new class of "flashlight cells," which came in fifteen different sizes.[23]

The standards tell us as much about the shortcomings of early primary cells as their properties. One major problem was unpredictable performance: the worst battery delivered 30 percent of the energy provided by the best battery. Another problem was storage: after six months, the service life of the average ignition battery fell by half. Some batteries were also prone to leaks, oozing electrolyte that could damage equipment. These problems were all exacerbated in warm temperatures. Although a warmer battery would yield better performance while in service, service life fell sharply when batteries were stored at warm temperatures. The US Bureau of Standards recommended storing and transporting batteries at 77°F or below. Because degradation was such a problem, the standards required that every battery sold be clearly labeled with a manufacture date or expiration date. For best results, batteries needed to be purchased by the "use by" date and kept cool.

The US Bureau of Standards worked with the battery industry to revise the standards for single-use batteries roughly once every five years. In 1927, the Bureau established a common nomenclature for batteries. It retained the common No. 6, and it standardized the sizes for the smaller flashlight batteries, which came in sizes labeled A, B, C, D, E, and F.[24] In 1937, the Bureau introduced standards for even smaller penlight batteries, including the now familiar AA and AAA sizes.[25] In each case, the dimensions of the batteries and the specified voltage match up with those on the market today. What has changed over time is the popularity of different battery sizes. As battery-powered devices became smaller and more efficient, smaller batteries became more common and larger batteries less so. In the 1920s, the No. 6 cell was most popular. In the 1950s, the D cell was the "standard" size. In the 1980s, AA batteries gained traction, powering the first portable cassette players, such as the Sony Walkman. Today, AAA batteries are most common. The large-format No. 6 cell disappeared in the 1970s.

The standards also document dramatic improvements in battery performance over time. Batteries manufactured in the mid-1930s provided significantly better service than did those tested in the late 1910s. Radio batteries lasted four times as long. Flashlight batteries lasted three times as long. Telephone batteries lasted twice as long. And the six-month degradation test fell too, dropping from a 25 percent loss in 1916 to a 7 percent loss in 1934. In short, the standards had made batteries a more predictable and reliable consumer product. Starting in 1930, the specifications also included minimum performance standards, which further ensured the predictability of portable batteries. As the standards noted, "The result of all these factors has been a considerable gain to the public at large."[26] Or, as a Ray-O-Vac battery company catalogue promised in the 1940s, portable batteries provided "Power When You Want It."[27]

Disposable batteries had emerged as a form of infrastructure that would help power the rise of a modern consumer electronics industry. Infrastructure usually refers to large-scale investments and systems, such as shipping canals, interstate highways, or, most relevant in this case, the electrical grid. What was fixed in the case of batteries was their size and their voltage. In the case of size, that was largely determined by human decisions, which were ratified by the standards committees. In the case of voltage, that was largely determined by the physical and chemical properties of the constituent materials and how they interacted in a battery.

Standardization drove a surge in consumer applications. Manufacturers began to design their new products around batteries, ensuring that flashlights, radios, cameras, and toys accepted the standardized sizes and voltages already available not just in the United States but worldwide. What made both this standardization and proliferation of batteries possible was a growing industry that harnessed resources from around the world to produce batteries by the billions. By World War II, no matter the brand of battery or where it was purchased, consumers could count on batteries to fit in their device and function with a reasonable degree of reliability—at least if it was not too warm or cold and the battery was not too old.

AN INCOMPLETE MATERIAL HISTORY
OF THE ZINC-CARBON BATTERY

In the past century, two chemistries dominated single-use batteries. In the late nineteenth century, battery manufacturers commercialized zinc-carbon batteries. After World War II, in response to concerns over the short shelf life and temperature sensitivity of zinc-carbon batteries, a Canadian inventor, Lewis Urry, invented the alkaline-manganese battery, which was similar to a mercury-alkaline cell used during World War II. Between the 1960s and 1980s, the higher-performing and more-reliable alkaline-manganese battery gained market share as a premium alternative to the zinc-carbon batteries. The downside to the alkaline-manganese batteries was that they contained approximately 1 percent mercury. In the late 1980s, battery manufacturers introduced the mercury-free alkaline-manganese battery, which eliminated the largest source of mercury in the nation's trash.

The provenance of the materials needed to manufacture batteries is as complicated as their chemistry. The story of these materials is complicated for two reasons. First, disposable batteries rely on multiple metals and chemicals. Although materials such as manganese, zinc, steel, carbon, graphite, ammonium chloride, and potassium hydroxide are essential battery materials, batteries have accounted for only a small fraction of the total consumption of most of those materials historically. Nor are the materials used in disposable batteries especially toxic (with the exception of mercury). Taken together, these factors mean the materials important to single-use batteries have drawn less scrutiny and left behind a shorter paper trail than other battery chemistries, notably the lead-acid battery.

What has distinguished the materials used in disposable batteries, however, is the importance of purity and consistency. Over the course of the twentieth century, improvements in battery performance and the transition to mercury-free batteries largely hinged on purer, more uniform, higher-quality materials. Today, disposable batteries have long shelf lives, perform with a high level of predictability, and are mercury-free. But achieving those qualities required investing a hundred times more energy in processing materials and manufacturing the batteries than they return in use.[28]

To begin to make sense of the material history of disposable batteries, I am going to focus on one material important to both the older zinc-carbon batteries and the alkaline-manganese batteries: manganese dioxide. By far the largest use of manganese historically and today has been as an alloy to strengthen other metals, such as steel and aluminum. Manganese alloys hardened steel railroad tracks and aluminum soda cans. Such uses accounted for more than 90 percent of manganese consumption in 1990.[29]

At the start of the twentieth century, however, the highest grades of manganese ore, specifically manganese dioxide ore or pyrolusite, were categorized as "battery ore." And for much of the twentieth century, demand for high-grade "natural" battery ore was high. That only changed when battery companies learned how to make "electrolytic manganese dioxide." It was a high-quality manganese dioxide synthesized from low-grade ores that became preferable even to its "natural" counterpart. It would become a staple of the modern battery industry, not just for disposable batteries but also for some lithium-based batteries.

Manganese dioxide is the biggest component of a disposable battery by weight. In manganese dioxide, the manganese atom's outer shell of electrons is less than half full. That makes it willing to accept additional electrons. In the case of a zinc-carbon battery, those electrons are supplied by a zinc anode. When connected by a circuit, the zinc anode gives up electrons, which the manganese dioxide happily accepts (while positive ions move through the electrolyte, balancing out the chemical reaction). The strength of that relationship—meaning the eagerness of electrons to leave the zinc and be taken up by the manganese dioxide—determines the battery's voltage.[30]

Manganese dioxide is not unique in its ability to accept electrons. Other metals, such as mercury or cobalt, have been used as the cathodes in other battery chemistries. Yet, since batteries were first commercialized in the early twentieth century, electrochemists have often preferred manganese dioxide for several reasons: it is an abundant material (it is the fifth-most-abundant metal in the Earth's crust); it is nontoxic (relative to other potential cathodes, such as mercury); and it is relatively inexpensive. By no means, however, did this make manganese dioxide a no-brainer for battery chemists. As two researchers commented in a 1952 review article, "It is doubtful if any single problem in dry cell technology has claimed as much attention as the selection of manganese dioxide for depolarizer use."[31]

One problem, these researchers explained, is that while manganese dioxide has a strong electrochemical desire for electrons, it is not very conductive. To help electrons find their way through the circuit and to the manganese dioxide, battery chemists added carbon to the battery's cathode. The carbon, most often in the form of graphite, creates a well-organized and dense atomic network that makes it easy for electrons to flow. But striking just the right proportion of manganese dioxide to graphite, optimizing the materials' surface areas and other physical properties, and eliminating impurities would take decades to accomplish.

Although manganese is an abundant mineral, the natural battery-grade ore had a well-defined historical geography. Prior to World War I, the Chiaturi mine in the Caucasus region of Russia, or the country of Georgia today, dominated the global supply of manganese. That meant when battery manufacturers needed manganese, they could tap into an existing trade. And the Chiaturi mine stood out because the mining was relatively easy and the quality of the manganese dioxide ore high. Early accounts describe seams of soft ore, easily mined along the summits of mountains above the Kvrilli River. Basic sorting and cleaning yielded an ore that was 86 percent manganese dioxide, less than 1 percent iron, with traces of copper, nickel, and bauxite. The ease of production and transport kept costs down and put Chiaturi at the center of the global manganese trade.[32] In the twentieth century, the region remained an important mining district, becoming famous for an extensive Stalin-era network of cable cars that carried miners between the mining towns and mountain mines.

World War I interrupted the supply of manganese from the Caucasus region. The Chiaturi mine itself had been German owned, and the war disrupted production at the mines, with U-boats threatening shipments to the United States from brokers who had stockpiled Russian ore in Britain. A small mining company in western Montana, a legacy of the region's late nineteenth-century silver boom and bust, saw an opportunity. As the Philipsburg Mining Company noted in 1916, "the matter of manganese ore is very interesting, and may prove profitable during the period of the war."[33] Remarkably, the small mining company's records are archived in Montana's historical museum in Helena. Those records give insight into how the Philipsburg Mining Company found itself at the center of the material economy of the American battery industry from the mid-1910s through the late 1920s.

Despite its promise, Montana manganese ore was not as high quality as that from Russia. In 1916, the Montana miners began trying to drum up business with battery manufacturers, most of whom were located in the Midwest. Initial shipments to battery manufacturers were low quality. The trouble was that the battery manufacturers were picky: they needed a high-grade ore, and they had little tolerance for impurities.[34] The "battery men" were also in urgent need of manganese dioxide. They made recommendations for how to crush, wash, and process the ore into battery-grade material.[35] Following their recommendations, the Philipsburg Mining Company broke ground on a concentrating mill in 1917 that would allow it to "turn these lower grade ores into a fine product."[36] By 1920, Montana supplied almost all of the nation's battery-grade manganese dioxide, amounting to twenty thousand tons per year.

Making the most of Montana manganese dioxide ore required changes for the battery manufacturers too. National Carbon Company made one of the most important advances in the early 1920s, as it adjusted its manufacturing to use the Philipsburg ore. Instead of grinding the manganese dioxide and carbon for the cathode separately, as had been the practice, they began grinding the materials together. It was a seemingly simple change, but the joint processing better coated the manganese dioxide with carbon, making it easier for the electrons to flow through the battery, which improved its conductivity and yielded "materially better" battery performance. Their preferred mixture started with "Philipsburg pyrolusite."[37] In 1952, electrochemists singled out that particular manufacturing change as the most important in improving battery performance in the first half of the twentieth century.[38]

Montana met much of the nation's demand for battery-grade manganese dioxide through the mid-1920s. During World War I, that helped keep soldiers supplied with batteries for early radios and flashlights. The real boom came after the war, however. Montana manganese dioxide helped power the nation's newfound fascination with the radio. Radio sales surged with the start of commercial broadcasting in 1920. For a few years, nearly every radio ran on batteries. As the radio helped stitch together the beginnings of American popular culture, that was made possible by shipments of electron-hungry manganese dioxide from the foothills of western Montana to battery manufacturers, mostly in the Midwest, who then shipped batteries to retailers around the country. In 1924, its peak year, Montana produced twenty-five thousand tons of battery-grade manganese dioxide—enough ore to manufacture hundreds of millions of No. 6, radio, and flashlight batteries.[39]

Montana's manganese boom ended as swiftly as it had arisen. Part of the problem was a limited supply of ore. In the early 1920s, Philipsburg opened up a new vein of manganese dioxide, but it came tainted with lead. Neither the miners nor the battery men noticed the lead, which measured less than 0.5 percent of the ore, until retailers across the country started complaining of faulty batteries.[40] The Montana mines began losing their market to imported ore. They competed with supplies from the Caucasus and a new manganese mine in present-day Ghana.[41] Although the United States adopted protective tariffs in 1922, those were not enough to protect the Montana mines. And, finally, a sharp downturn in demand for batteries hit the industry and the Montana manganese miners hard. Starting in the mid-1920s, new "socket-powered" radios, which plugged into light outlets, weakened demand for batteries. Battery sales fell by nearly half between 1926 and 1928. Manganese production from Montana fell to one-quarter of its 1924 peak and dwindled through the 1930s and 1940s.[42]

During the Montana mines' heyday, much the manganese ore was shipped to Madison, Wisconsin, home of the French Battery Company, which was known for its Ray-O-Vac batteries. The French Battery Company had grown since its founding in 1906 to include two factories and two thousand employees by 1928. Yet, its annual reports, the company's archives, and other newspaper coverage reveal little about what happened inside its factories.[43] One memo dated from 1924 reveals that state health officials required the French Battery Company to improve its ventilation system to protect workers, but there is little information about occupational safety or workers more generally. In fact, it is much easier to track the success of the French Battery Company's team in Madison's industrial baseball league. Their chief rival was the crosstown Burgess Battery Company.

The story that the French Battery Company's limited archives and the newspapers do emphasize is one of a successful and growing industry that depended on high-quality raw materials from across the country. In its 1924 *Sales Idea Book* distributed to retailers, the company highlighted the "control section of its laboratory," which analyzed "each and every raw material," starting with samples of ores shipped direct from mines or suppliers. This process, the company promised, guarded against "any defects due to impurities or poisons."[44] And that was how the company advertised its batteries to customers. It emphasized its long experience, ingenuity, and supply of high-quality materials. The result, it promised, was a "perfect product."[45]

In truth, however, these advances did little to distinguish the French Battery Company from its leading competitors. Across the industry, a driving focus had been securing and improving supplies of raw materials. When the government released the 1937 update to battery standards, it reported industry-wide improvement in battery life. To give but two examples: D-size flashlight batteries lasted three times as long as they had in 1910. Radio batteries lasted nearly five times as long as they had in 1918.[46] Few companies staked their claim on the quality of their raw material supply by the 1930s—gone were long, wordy advertisements explaining raw materials and detailing quality-control departments. Instead, they began trading on their brand names and snappy slogans—a sign of the rise of a modern consumer culture, of which batteries would be an important part.

Batteries were in short supply during World War II. The shortages drew the most public attention in rural communities, where a shortage of batteries meant a shortage of wartime information. "Probably never before has radio service meant so much to farmers as now, with the whole world depending on American farm production, and with the possible outcome of the war hinging on it." In the case of radios, listening hours had "jumped three to five hours" daily.[47] As part of the war effort, listeners were encouraged to ration their listening to save batteries.

Conserving batteries at home meant more batteries to meet the needs of soldiers. And during World War II, batteries had become "indispensable" on the battle front—they powered flashlights, portable radios, walkie-talkies, mine detectors, and bazookas. The army relied on batteries to help power more than seven hundred pieces of war matériel.[48] The newly invented bazooka, for instance, relied on two C batteries to fire a rocket.[49] An American general described the new backpack-size "walkie-talkie" as "exactly what is needed for frontline communications." The problem, he explained, was "keeping them supplied with fresh batteries."[50]

Demand for batteries surged during the war. Production for the military more than doubled to 1.5 billion batteries from 1943 to 1944, and even that was not enough.[51] The squeeze on batteries at home was not just a result of increased military demand, however. It was also a product of the military's need for a better battery. Despite the improvements in zinc-carbon batteries, they still performed poorly in warm and cold conditions, were susceptible to moisture, and remained highly perishable—the longer they were stored, the worse they performed. This became an urgent problem for the military as operations moved into warm and humid conditions in the Pacific theater in 1942.

To solve this problem, the US Signal Corps turned to Samuel Ruben, an independent New York–based inventor who was later described as the "Wizard of New Rochelle."[52] Ruben, who never earned a college degree, ran a small laboratory that supported a remarkably productive career. Ruben claimed three hundred patents, earned millions of dollars in royalties, and won numerous honors and awards. Among his inventions was a new battery chemistry that better withstood high temperatures and humidity and delivered four to seven times more energy than a conventional dry cell battery. Explained most simply, Ruben's cell, invented in 1942, relied on a mercuric oxide cathode (mixed with graphite), a zinc anode, and a potassium hydrox-ide electrolyte, all sealed in an airtight steel container. Mercuric oxide shared manganese dioxide's affinity for electrons, but it was much more stable. The airtight steel canister protected it from moisture. Because of its ability to with-stand heat and humidity, the military described it as the "tropical battery."[53]

The performance of Ruben's new battery was so promising and the need for a better battery so urgent that the US Signal Corps acted rapidly. First, it classified the mercury cell as confidential, delaying a patent application until the end of the war. Second, it looked to the battery industry, already on a wartime footing, to put the new battery into production immediately. Indeed, the national battery shortage was in part a consequence of battery manufacturers retooling their operations to produce the new battery. By 1945, P. R. Mallory, which began marketing its batteries using the Duracell name in 1965, and Ray-O-Vac were producing one million cells daily.[54] That meant as US soldiers advanced in the Pacific, taking Saipan, the Philippines, and Iwo Jima, they relied on Ruben's tropical battery—invented fewer than three years before—to power their military gear. As one government official put it, "The mercury battery saved the day."[55]

Wartime demand for batteries—especially the new mercury-based battery—served as a sharp reminder of just how dependent the United States remained on foreign supplies of key raw materials. Throughout the war, sup-plies of battery-grade manganese dioxide and graphite had been a constant concern. Manganese was imported from West Africa, and graphite con-sumption for nonbattery uses was curtailed. But no one was prepared for the surge in demand for mercury. As the government reported in 1944, the new tropical battery created a demand "so large as to reverse completely the out-look for the mercury industry."[56] Early in the war, the United States had ramped up domestic production to meet its limited mercury needs. But domestic production could not match the demand for Ruben's battery.[57] By May 1945, as production reached full tilt, the United States' mercury demand jumped to five times what it had been the year before. Government stockpiles helped meet some demand. But most of the mercury was imported from Peru, Chile, Canada, Mexico, and, especially, Spain.[58]

Ruben's autobiography recounts the challenges in scaling up produc-tion: limited materials, manpower shortages, manufacturing difficulties, and worker safety. Of those, he likely underestimated the risks to workers. At battery factories in Wisconsin and Indiana, public health officials docu-mented the challenges of protecting workers. In 1945, the US Public Health

Service concluded that mercury vapor concentrations were "well above maximum allowable concentration." In one plant, sixty of eighty workers servicing the cell aging room exhibited signs of mercury poisoning. At another plant, urine samples revealed mercury concentrations that exceeded present-day health standards by a factor of ten.[59] Although Ray-O-Vac and Mallory worked to improve working conditions, relying on personal respirators and better ventilation, producing the tropical battery meant factory workers faced their own risks on the home front.

When word began to spread about Ruben's new battery in 1945, it was described as a "miracle battery" that had done a "big job in the tropics."[60] That advertising was important, as the military canceled its contracts at the end of World War II. Yet, companies like Mallory and Ray-O-Vac had only limited hopes for the mercury battery. They knew they faced a challenge: the new battery was much more expensive than the zinc-carbon batteries. That limited sales of the new battery to specialty applications, such as watches and hearing aids. But, as the historian Eric Hintz has explained, Ruben's invention did more than just meet wartime demand. It also transformed the battery industry. The mercury battery hastened the trend toward a new, sealed alkaline-based battery chemistry and a new generation of portable electronic devices that would explode in popularity in the post–World War II era.[61]

ELECTRIFYING THE RACE FOR RAW MATERIALS

Today, the global trade in materials rarely draws much public attention. When it does, attention usually centers on foreign trade in oil. There are exceptions. In the early 2000s, concern about how the trade in tantalum, tin, and tungsten was being used to finance armed conflicts in Central Africa resulted in reporting around "conflict minerals." In the early 2010s, China clamped down on exports of rare earth metals that were vital for many advanced technologies from smartphones and hybrid cars to wind turbines and weapons systems. That sparked renewed concern about the United States' dependence on foreign supplies of critical minerals important to a clean energy economy and national security.[62]

But during and after World War II, the United States' growing dependence on foreign supplies of raw materials was a pressing concern for policymakers and the public. With only 7 percent of the world's population, the United States already consumed 50 percent of the world's minerals and 70 percent of its oil in 1950.[63] The question, as the director of the Bureau of Mines put it, was, how long can we "maintain this rate of increased production and consumption without running out of raw materials"?[64]

Although the United States had abundant supplies of coal, iron ore, and phosphate, the Bureau of Mines warned that was not the case for many other materials, several of which were important to the battery industry. It warned that the United States could exhaust its reserves of mercury and manganese by the early 1950s, lead by the early 1960s, and zinc and copper by the early 1970s. With the material challenges of World War II still fresh and new conflicts emerging on the Korean peninsula, securing raw material supplies became a matter of national security—not only to ensure soldiers were well equipped but also to sustain the economic growth and political might of the "free world" at the start of the Cold War.[65]

These broader concerns culminated with the publication of what was known as the Paley Report in 1951. Commissioned by President Truman, the Paley Commission's chief concern was securing the nation's supply of resources, not only to sustain the American economy but to fend off the "new Dark Age" of communism.[66] The report's cover was telling: it depicted the globe defined not by nation-states or political boundaries but by oil derricks, ore piles, and other natural resources. As the historian Megan Black explains: "In this rendering of the globe, natural resources dominate, while national sovereignty falls from view. Only gridlines of longitude and latitude remain— manmade markers that rather than dividing peoples into nations, divided nature into units awaiting utilization."[67]

Considering the history of the battery industry reframes the scramble for resources at the start of the Cold War era in an important way. The most urgent concern for the battery industry was not just securing supplies of raw materials—which were in short supply domestically—but developing the technologies necessary to process those raw materials into the high-quality materials needed for batteries. This problem was especially true for manganese dioxide, which made up the largest proportion of the disposable zinc-carbon batteries. Today, as the United States plans for a clean energy future, this challenge is just as urgent. Not only does the United States lack many of the raw materials most important to manufacturing lithium-ion batteries, but it also lacks the processing facilities needed to turn lithium, nickel, cobalt, and graphite into the high-quality precursors needed to manufacture the anodes and cathodes required for lithium-ion batteries. Then and today, what counts as a "raw material" depends as much on advances in processing capacity as it does on finding new ore bodies.

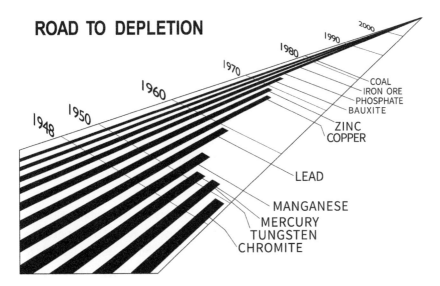

ROAD TO DEPLETION

COAL
IRON ORE
PHOSPHATE
BAUXITE

ZINC
COPPER

LEAD

MANGANESE
MERCURY
TUNGSTEN
CHROMITE

FIGURE 2.2 Concerns about resource scarcity mounted after World War II. The Bureau of Mines warned that the United States was on the road to depletion for key minerals. Mineral Resources Development, US Senate Hearing, 88th Cong., 1st Sess. (July 1949).

The wave of independence that swept through the European empires in the wake of World War II took root early in West Africa, destabilizing the global supply of manganese. The Gold Coast had supplied some manganese for the steel industry and almost all the manganese for the battery industry. That drew the US State Department's attention. One official reminded his colleagues of the "absolute necessity" of African manganese in 1948.[68] In the late 1940s and early 1950s, the Department of State engaged in discussions with the African Manganese Company in the hope of securing an expanded supply from what became the independent nation of Ghana. But those overtures yielded little progress, which only heightened US concerns over its dependence on imports of critical raw materials.[69]

One company that had boomed with the wartime demand was Ray-O-Vac. During World War II, Ray-O-Vac had relied on imports of manganese dioxide from the Gold Coast. But, as sales of its leakproof battery grew in the postwar era, its self-described efforts to "bird-dog" the globe in search of the stores of battery-grade manganese that "nature [had] distributed helter skelter" increasingly came up short. In 1951, the company warned it was "seriously handicapped" by an unreliable supply of high-quality manganese dioxide, especially with growing military demand during the Korean War.[70]

Congress convened the Strategic and Critical Minerals and Metals Hearings in the spring of 1948. One committee member urged the government to stockpile all the manganese ore it could from foreign sources. That ore was "more valuable to the Nation than your gold at Fort Knox."[71] But most of the hearings centered on how lower-grade manganese ores, both domestic and foreign, could be turned into valuable raw materials. In the 1940s, the manganese industry had begun to scale up a new process for converting ore into metal. It was an electrolytic process—already in use in other metal industries, such as copper and aluminum—that relied not on heat but on electricity.

It is worth taking a brief detour from Congress to learn about electrometallurgy as a form of metals processing, not just because of its importance to expanding the supply of manganese in the 1950s, but because it is likely to play a key role in advancing more sustainable metal refining processes in the twenty-first century. Electrometallurgy deployed the same electrochemical principles that underpinned the functioning of a battery to a different end. Whereas a battery harnessed a chemical reaction to generate electricity, the processes of electrowinning and electrorefining relied on electricity to drive a chemical reaction that yielded purer materials. In the case of manganese dioxide, after leaching manganese from a crushed ore with sulfuric acid and passing electricity through the manganese-rich electrolyte, it was possible to harvest extremely pure manganese dioxide that plated on the electrodes. That made it possible to refine manganese ores into a high-quality manganese dioxide that was even purer than "natural" battery-grade ore.

The electrolytic process had two important advantages that benefited the mining companies and the battery industry. First, the electrolytic process could produce battery-grade manganese dioxide from lower grades of ore. That dramatically expanded what counted as a potential reserve of battery-grade manganese ore. It required more processing and energy, but the result was 99.99 percent pure. That meant, second, electrolytic manganese dioxide yielded an even longer-lasting battery. As a result, in the years after World War II, the battery industry began to look more broadly around the globe for ore deposits suitable for producing electrolytic manganese dioxide, and companies like Ray-O-Vac and Union Carbide invested in new electrolytic plants to turn that ore into synthetic manganese dioxide.[72]

Starting in the 1950s, the scale of electrolytic manganese dioxide production began to expand, primarily in the United States and Japan. The industrial geography of production was telling. In the United States, the new plants were located in areas with abundant and inexpensive hydroelectricity, such as Tennessee, Oregon, and Nevada. Although most electrolytic processes require high-temperature roasting at an initial stage, cheap electricity made it economical to process low-grade ores into high-quality battery-grade manganese dioxide.[73] Ray-O-Vac located its first plant in Salem, Oregon, drawing on hydroelectricity from the Columbia River. Another plant was started in Henderson, Nevada, in 1951, powered by electricity from the Colorado River. The Nevada site has been in production ever since.[74]

Electrolytic manganese dioxide's usage grew steadily in the postwar era. Its high purity resulted in a zinc-carbon battery that lasted twice as long as those using natural manganese ore.[75] But that purity and performance came at a cost. It required at least twenty times more electricity to process the electrolytic manganese dioxide alone than the battery returned during use.[76] And, as one insider's 1987 survey of the industry noted, it resulted in "vast amount[s] of waste and effluents," demanding careful treatment.[77] Yet, as demand for reliable, longer-lasting batteries continued to grow, these challenges did little to slow the industry in its quest for longer-lived and more reliable batteries.

The little-known story of electrolytic manganese dioxide, which played such a pivotal role in making for a better battery in the postwar era, marked a much larger turning point in material history. I have always envisioned metals production as a process centered around an enormous furnace, plumes of smoke, and vats of glowing metal. In fact, an activity in a colleague's environmental science courses is to smelt a fist-size nodule of copper ore into pure copper. It is an exhausting project, requiring students to pump a set of manual bellows for several hours in an effort to reach and maintain temperatures above 1,000°C. After all that work, if all goes to plan, the result is a pea-size hunk of copper.

The story of twentieth-century metals production, however, has increasingly turned not just on pyrometallurgical processes but on the kinds of electrometallurgical and hydrometallurgical processes important to the production of electrolytic manganese dioxide. The best example of this is aluminum, which has been produced through electrolytic processes since the late nineteenth century. Today, 90 percent of zinc is produced electrolytically. As important as these processes have been in the past, they are likely to be even more important in the future. Unlike pyrometallurgical processes, which usually depend upon fossil fuels to achieve the high temperatures needed for smelting, electrometallurgical and hydrometallurgical processes can be powered by electricity from renewable energy sources, making them potentially more sustainable.

DOING MORE WITH LESS: TRANSISTORS

Sales of batteries exploded in the 1950s. As *Popular Science* magazine noted, "A lot has happened since the days when flashlight batteries were used for flashlights." The so-called flashlight batteries were being used to power everything from portable record players to cocktail mixers to movie cameras to dictating machines to kitchen wall clocks. *Popular Science* marveled at what could now be done "without plugging in an electric cord."[78] With so many uses, one commenter observed, it seemed everyone was "in the big-time battery buying business."[79]

The industry's annual sales topped one billion batteries, or approximately six batteries per American in the 1950s. Not only did the number of batteries sold increase, but the value of those batteries increased too. Smaller, better-performing batteries to power new transistor radios were gaining market share, led by C- and AA-sized batteries. Ray-O-Vac described 1961 as a "red-letter year" for the company.[80] Within two decades, however, Duracell and Energizer pushed Ray-O-Vac to the sidelines, as these rising titans of the American consumer industry and throwaway culture captured the market for a new generation of better-performing alkaline manganese batteries.[81]

In the late 1950s, electrochemists at Duracell and Energizer each developed versions of the tropical battery that replaced the mercuric oxide cathode with manganese dioxide—a new chemistry made possible by the highly pure and refined qualities of the electrolytic manganese dioxide. The new alkaline-manganese batteries had many of the advantages of the tropical battery—they were more powerful, temperature resistant, and stored well—and they were also cheaper. And new high-drain devices, such as battery-powered cassette players and camera flashes drove demand for higher-quality batteries. Both Energizer and Duracell brought alkaline-manganese batteries to market in the late 1950s and early 1960s. It was this battery that first gained Duracell's copper top (on the positive electrode), which became a company trademark.

The other major technological shift in the alkaline-manganese battery happened on the other side of the battery: at its negative terminal, the anode.

Instead of relying on a zinc case, which had doubled as the battery's anode since the start of the twentieth century, the new alkaline-manganese batteries used a powdered zinc metal for the anode. From an electrochemical perspective, powdered zinc was advantageous because it dramatically increased the surface area of zinc exposed to the electrolyte. That increased its reactivity, which made it a better source of electrons, which improved performance. It also made it possible to pack the anode and cathode inside a strong and durable steel can, which made the new alkaline-manganese batteries more durable and less likely to leak.

The advent of the alkaline-manganese dioxide battery marked a little-recognized waypoint for the battery industry. As late as the 1950s, Ray-O-Vac had prided itself on "bird-dogging" the world for high-quality raw materials.[82] Although such information was never well disclosed, earlier batteries depended on high-quality raw materials sourced from specific locations that had favorable geological deposits—especially manganese dioxide. But the high levels of purity resulting from electrolytic processing diminished the importance of where materials were sourced from and increased the importance of how they were processed. Indeed, many key materials for the new alkaline-manganese batteries—the manganese dioxide, zinc, and the potassium hydroxide electrolyte—depended upon the high degree of purity resulting from electrolytic refining. And in the latter half of the twentieth century, improvements in alkaline batteries rested almost entirely on more sophisticated processes for purifying and improving these materials.

Despite the advantages conferred by powdered zinc's large surface area, it also posed a serious challenge. Although powdered zinc—which was produced electrolytically like the manganese dioxide—was highly pure, impurities such as iron and lead remained. Even though the impurities only measured in the parts per million, the impurities could still cause unwanted reactions inside the battery, which reduced battery performance and could result in swelling and leaking. To solve this problem, the powdered zinc was combined with mercury. That served two purposes. First, it inhibited the unwanted reactions in the zinc anode. As one battery engineer explained, the mercury just served to "level everything out."[83] Second, adding mercury to the anode made it easier to pump the zinc-electrolyte mixture into the steel can during manufacturing.[84] And in the 1960s, when consumer products had few labeling requirements nor were there any disclosure laws, few people knew that every new alkaline-manganese battery put on the market—and later thrown away—contained mercury.

What companies like Duracell did advertise, however, was the performance of the new alkaline-manganese batteries. In 1970, Duracell asked, "Will you buy an old-style battery" that had been "unchanged since 1888"? Or will you buy a Duracell? It described its new batteries as being "as different from an ordinary battery as your car is from a horseless carriage."[85] Tests in 1978 demonstrated that the batteries had two to five times the capacity of the zinc-carbon batteries they replaced.[86] For that reason, consumers turned to alkaline batteries to power a growing array of portable electronics. Many of those were premium products, for which companies like Duracell and Energizer offered a premium battery. The introduction of the Sony Walkman, the first personal audio player, exemplified this trend. It was a battery-powered cassette player with lightweight headphones that put convenience and portability ahead of audio fidelity. The result is what is widely credited as the first breakout consumer electronic device. Five years after its 1979 introduction, consumers had purchased ten million Walkman players, and competitors had flooded the market with dozens of competing devices.[87]

In an age of smartphones, it is a wonder what a storm of cultural handwringing the advent of the Sony Walkman provoked. Much of that consternation centered around the ability of individuals, sporting telltale headphones, to at once be physically present in communal spaces—on the street, in the subway, or in the restaurant—yet absorbed in their own individualized soundscape. Unlike the transistor radios that preceded them, the Sony Walkman catered to its listener. You could listen to whatever you wanted. "The Walkman is replacing certain drugs as a mind- and mood-altering device," warned a popular newspaper columnist in 1981.[88] It played into growing concerns in the early 1980s about individualism, hedonism, and self-interest that would reshape American culture in the late twentieth century. As the Walkman's most recent chronicler, Rebecca Tuhus-Dubrow, explains, it introduced the ideas of multitasking, always-on personal entertainment, and the "notion of electronic gadgets as accessories."[89] All the Walkman required was a cassette tape and fresh set of AA batteries.

The Walkman became notorious for "gobbling batteries." One New York resident reported using more than a hundred AA batteries a year to keep his Walkman running.[90] He was not alone, nor was it just Walkmans that were gobbling up batteries. By the early 1990s, disposable battery sales in the United States had reached four billion per year. That made disposable batteries a growing part of a torrent of trash.[91] A profusion of products, many designed to be convenient and disposable, meant consumers were buying and throwing away more and more stuff. Although batteries accounted for less than one-tenth of the nation's household trash, batteries accounted for the vast majority of the toxic heavy metals in household waste.

TOXIC BATTERIES AND A NATIONAL TRASH CRISIS

As concerns about trash mounted, the federal government began sticking its regulatory nose into the nation's trash bins. Without action, one congressman warned in 1976, "we may be faced with using our parks as disposal sites in the not-too-distant future."[92] Federal laws such as the Resource Conservation and Recovery Act of 1976 aimed to limit trash dumps, regulate hazardous waste disposal, and promote resource recovery. The law did help jump-start curbside recycling programs, which collected materials such as paper, glass, and plastics. It also played an important role in cleaning up the lead-acid battery industry. But the most immediate consequence was a wave of landfill closures around the country—3,500 landfills closed down between 1979 and 1986, largely because they could not comply with regulations banning open dumping and managing polluted runoff. That forced cities to look for other ways to get rid of their trash: shipping it across state lines, exporting it to other countries, or burning it in more advanced waste-to-energy incinerators (a strategy that had largely been abandoned out of concern for air pollution but was regaining attention in the 1980s for its potential as an energy source).[93]

Concerns about the toxicity of disposable batteries grew in the 1980s, especially in countries such as Japan, Sweden, and Germany, where trash incineration remained common. A 1984 *New York Times* article reported on "fears that Japan is slowly being contaminated by the dry cells that power its calculators, cameras, portable stereos, and watches."[94] Driving these concerns was the transition from the old zinc-carbon to the new alkaline-manganese batteries. Each alkaline battery contained one hundred times more mercury than the old zinc-carbon batteries they replaced. Environmental activists warned that all those "Coppertops, Ultralifes, and Energizers" posed two threats: if burned, they were a source of toxic air pollution; if landfilled, the metals could eventually leach into groundwater.[95] A 1986 study concluded that the best way to control heavy metals in the nation's household trash was to target disposable batteries—the "obvious" target—either by recycling them or banning the use of heavy metals.[96]

As the United States began to face up to the deluge of trash in the 1980s, Minnesota emerged as a policy incubator. In southeastern Minnesota, the state's porous limestone geology was especially prone to groundwater contamination. That made siting landfills and managing polluted runoff especially problematic. To reduce landfill use, the City of Minneapolis decided to build a European-style waste-to-energy facility downtown in the late 1980s.[97] Instead of paying to export garbage to distant landfills, trash could be handled locally, generating electricity and revenue instead.[98] Although the incinerator would have pollution-control technology, reducing emissions from heavy metals remained a challenge. That posed the greatest threat to children living nearby in Minneapolis and St. Paul, who were already presenting with high lead levels.[99] There was also concern that mercury from the facility would further contaminate Minnesota's lakes, where high levels of mercury had already contaminated fish and prompted health advisories.[100]

The proposed waste-to-energy facility put disposable batteries at the center of Minnesota's waste management regulatory crosshairs in the late 1980s. Instead of classifying all those AA, C, and D batteries as household trash, as they were most everywhere else in the country, Minnesota began considering policies to reclassify them as toxic waste requiring labeling, a deposit-refund system, and special handling, either through hazardous waste processing or recycling.[101] Those proposals drew the enthusiasm of environmental groups. Why, they asked, when it came to recycling, were "batteries not included"?[102]

The industry had an answer. Based on their studies in the 1980s, batteries posed little threat to human health and did not warrant special treatment. The last thing the industry wanted were regulations that singled out batteries. A deposit-refund system would be cumbersome to implement (no one wanted to be counting returned batteries at the cash register when buying new batteries). And forcing manufacturers to recycle batteries, warned an industry lobbyist, was a quick way to make them go "broke."[103] The Dry Battery Section of the industry's trade group, the National Electrical Manufacturers Association, maintained that disposable batteries were safe for disposal, whether landfilled or burned.

The battery industry knew this was a problem it had to confront head on. With environmentalists focusing more attention on the need to recycle batteries, awareness of the threats of heavy metals pollution mounting, and regulatory pressure growing in Minnesota (and other states such as New York, New Jersey, and California, and especially the European Union) the industry accelerated its efforts to phase mercury out of its batteries.[104] If alkaline batteries contained no mercury, that undercut the strongest argument for giving batteries special treatment.

Battery manufacturers could eliminate mercury. The challenge was doing so without compromising battery performance or stability. Taking that step was much easier if it was required of all manufacturers simultaneously. In 1989, the battery industry agreed with the State of Minnesota to a scheduled phase out of mercury in alkaline batteries, but it strongly opposed any deposits, take-back policies, or recycling mandates for the common zinc-carbon and alkaline-manganese batteries. When it was passed in 1990, Minnesota's was the first battery law in the country that required manufacturers to reduce mercury. It limited mercury levels to 0.025 percent in disposable batteries by February 1, 1992, and required collection of more toxic batteries. In turn, the legislation omitted requirements for collecting alkaline batteries or recycling them.[105]

Minnesota's law had national consequences. "We're on the cutting edge," explained its chief sponsor. "We will have the highest standards in the United States."[106] Battery manufacturers began to market mercury-free batteries nationally, and other states began to adopt similar policies, including New Jersey and New York. The battery bill put Minnesota at the vanguard of regulations that placed new responsibility on industry for reducing the hazards of their products at end of life. Its mandate to reduce the mercury content of disposable batteries was an example of "source reduction policy." Its mandate requiring the collection of more toxic batteries was one of the earliest examples of an "extended producer responsibility" policy, which made industry—instead of municipalities or taxpayers—responsible for managing products at end of life. Both would become increasingly important models for policy action in the 1990s and 2000s, especially as public attention turned toward the management of old computers, monitors, and fax machines—what we now lump together as e-waste.[107] By the mid-1990s, thirteen states had passed similar battery laws, and in 1996, Congress enacted the Mercury-Containing and Rechargeable Battery Management Act, which codified the ban on mercury in disposable batteries nationwide.[108]

What is most remarkable about the transition to mercury-free alkaline batteries is how quickly it took place. By the early 1990s, all major manufacturers had transitioned to mercury-free alkaline batteries. Eliminating the mercury put significant pressure on the zinc industry. To eliminate mercury, battery manufacturers needed higher-quality powdered zinc—well beyond commodity-grade zinc—doped with proprietary additives that inhibited gassing. Additional electrowinning helped produce a purer zinc powder, containing only a few parts per million of iron, which was the most problematic impurity. (Remember, the earliest "natural battery ore" from the Caucasus was prized because it was less than 1 percent iron.) That allowed for a reduction in the mercury in the late 1980s. Eliminating mercury entirely in the early 1990s, however, required the addition of other impurities measured in the parts per million, such as lithium, indium, bismuth, or aluminum. Indeed, each manufacturer developed a proprietary strategy. The race to eliminate mercury had jump-started a surge of innovation in the battery industry that would last another decade.[109]

In retrospect, the elimination of mercury from alkaline-manganese batteries is a little-known, but important, environmental and regulatory success story. This was how environmental policy was meant to work. As concerns about the potential consequences of heavy metals pollution in the municipal waste stream mounted, policymakers and industry responded. A cascade of policy action, which started in Europe in the mid-1980s and quickly caught on in states such as Minnesota and New Jersey in the late 1980s, led to congressional action in the United States that provided consistent regulations nationwide. It may be that the magnitude of the threat from batteries was never great—a position that the industry maintained. But it also became evident that the threat of regulations, such as mandatory collection or recycling, was the nudge the industry needed to invest in the research and development necessary to bring mercury-free batteries to the market. The results were unequivocal. By 1995, the use of mercury in batteries had fallen more than 99 percent from a decade earlier and levels of mercury in the waste stream fell too.[110]

THE COMPLICATED CASE FOR RECYCLING

The push to recycle alkaline batteries did not go away. As one battery industry executive acknowledged, even though mercury was gone, consumers still thought of batteries as "hazardous chemical plants." Although the industry described batteries as being made up of "natural" materials that posed no threat in the waste stream, they knew that when people went to throw batteries away, they paused, asking themselves, "Should I be doing this?"[111] The industry's official answer was yes, but internally the industry knew that many consumers felt guilty about throwing out batteries. And for a large-scale consumer-product industry, consumer guilt is cause for worry. The industry maintained that it was safe to throw out disposable batteries and that recycling was unnecessary through the early 2000s. That changed in 2011. In the face of growing pressure over the management of batteries, Energizer, Duracell, Rayovac, Panasonic, and Kodak all formed the Corporation for Battery Recycling to spearhead a new initiative: "a nationwide system that maximizes re-use of spent batteries with a goal of zero waste."[112]

To understand this turnabout requires a quick detour back into hazardous waste policy. In 1976, the Resource Conservation and Recovery Act (RCRA) aimed to improve management of hazardous waste nationally.[113] It established strict protocols for collecting, transporting, and processing municipal and hazardous wastes. But the law was written so broadly it had an unintentional consequence: it made it harder to collect and manage common consumer products, such as fluorescent lightbulbs, thermostats, and batteries, which contained small amounts of hazardous materials and posed a lesser threat to public health. Although it was better to keep those items out of the trash, the fear of getting tripped up by the RCRA's stringent requirements deterred municipalities, retailers, and others from collecting and properly managing these wastes. To solve this problem, the Environmental Protection Agency (EPA) issued the Universal Waste Rule in 1995, which relaxed the regulatory requirements for these common wastes, with the goal of making it easier to collect, transport, and recycle them. Most importantly for the battery industry, the new regulations excluded common household batteries from hazardous waste regulations—that made it legal to dispose of AA, AAA, and 9-volt batteries.[114]

But, as is often the case when it comes to environmental policy, what California does matters as much as what the federal government does. Although alkaline batteries passed the EPA's toxicity tests, which exempted them from RCRA, California regulators came to a different conclusion. Starting in the 1990s, California determined that alkaline batteries required special handling because of their corrosivity—the potassium hydroxide electrolyte in alkaline batteries is a strong base that can burn skin on contact or contaminate soil. Considering that batteries still made up a vanishingly small fraction of the waste stream, were unlikely to leach, and were hardly the only corrosive material in it (bleach is a strong base too), the industry maintained that California's policies were unnecessary.[115] But state environmental officials argued that the battery electrolyte could mobilize other toxins in municipal solid waste, thereby posing a public health threat.

Despite California's concerns and regulatory determination, in practice, nothing changed. Most batteries still wound up in the trash in California, as they did elsewhere. The industry tried to push a law exempting disposable alkaline batteries from California's hazardous waste regulations in 2000, only to have the bill vetoed by California governor Gray Davis on grounds that batteries could pose a threat to public health and were potentially recyclable.[116] The result was that starting in 2006, when the state fully implemented its version of the Universal Waste Rule, it became illegal to throw away disposable batteries in California.[117]

That put California at the forefront of the campaign to stop discarding and start recycling disposable batteries in the 2000s. California was not alone. The European Union had adopted regulations requiring collection of all batteries, with the goal of recycling them, in 2006. And other states and Canadian provinces were moving in the same direction. Even if recycling made environmental sense—which, it turns out, was an open question—the most immediate challenge was the cost. Disposing of batteries in municipal trash was effectively free, involving no additional costs. Sending a fifty-five-gallon barrel of alkaline batteries for hazardous waste disposal cost $159— not to mention staffing local collection and recycling facilities. Recycling the same fifty-five-gallon barrel of alkaline batteries cost $1,250.[118]

Why should cities and towns foot the bill for handling such waste when it was the battery manufacturers and retailers that profited from selling the batteries in the first place? That was a question that had implications that went far beyond disposable batteries. To answer it, environmental advocates and state legislators began pushing retailers to take responsibility for managing problem wastes, such as mattresses, mercury-containing thermostats, and electronic waste at end of life. The reasoning behind such "extended producer responsibility" laws was that if a product, such as batteries, required special treatment at disposal, then the manufacturers should pay for it. Extended producer responsibility legislation for batteries has been introduced to the California legislature almost every year since 2006.

This brings us full circle, back to the start of this chapter. It was around 2012 that I first tried to figure out what happened when I recycled my old AA and 9-volt batteries. I placed my batteries in the Big Green Box in my college's library. The library had purchased the box for around sixty dollars. "It's Simple: Buy it. Fill it. Ship it."[119] It included the sturdy cardboard collection box, eco-friendly messages, and a prepaid shipping label. Once it was full of batteries, United Parcel Service picked up the box and shipped it to Anaheim, California. The Big Green Box was an offshoot of a California company, Kinsbursky Brothers, that had started as a scrap yard, expanded to lead-acid battery recycling, and then expanded again in response to California's restrictions on battery disposal.

Once my spent batteries arrived in Anaheim (nearly three thousand miles away from Massachusetts), they were manually sorted by type.[120] The box likely contained mostly disposable alkaline batteries, but those had to be separated from any lithium, nickel-cadmium, or small lead-acid batteries, each of which required a different recycling process at a different facility (and posed a significant hazard if accidentally fed into the wrong recycling system). Once my disposable batteries had been sorted, they were shipped out again to the actual recycling facility.

No matter where they wound up, however, it was highly unlikely that my batteries became feedstock for new alkaline batteries. Instead, the materials were recovered to varying degrees for other purposes. An MIT study commissioned by the National Electrical Manufacturers Association analyzed existing and potential recycling options in 2010. Broadly speaking, the existing strategies fell into two categories: Pyrometallurgical strategies heated the batteries with other nonbattery scrap materials to more than 2,000°F and recovered metallic zinc and steel, with lower-quality manganese and metal scrap going into slag for cement or asphalt. Mechanical strategies ground the batteries into a powder and used magnets to recover steel, hydrometallurgy to recover zinc and manganese, and combustion to recover energy from plastic and paper.[121] In both cases, the recycling processes resulted in a form of downcycling—high-quality raw materials became lower-quality secondary materials.

My spent battery's long journey from Massachusetts to California revealed a basic problem with battery recycling: it often took more energy to collect, aggregate, ship, sort, and recycle the batteries than was actually saved by avoiding the production of new raw materials. And much of that energy expenditure was in collecting and transporting the batteries. This was a point that the battery industry had been making since the 1980s: in the case of disposable batteries, not only was it costlier to recycle batteries, but it was also more resource intensive than simply throwing batteries in the trash.[122] But it was also a claim the industry wanted to test, in light of the growing pressure to collect and recycle batteries.

That was the reason the industry commissioned the MIT researchers to conduct a life-cycle analysis of single-use batteries—comprehensively evaluating the life cycle of batteries, from mining raw materials to manufacturing to the costs and benefits of existing and potential recycling scenarios. The results were complicated. In the case of energy consumption, greenhouse gas emissions, and resource consumption, recycling was more costly than landfilling for existing recycling scenarios. Where it was a net positive was in reducing toxicity—meaning the threat of pollution to human health and the environment—although this was also the metric with the highest levels of uncertainty.[123]

One way to read these results was that recycling made little sense, either environmentally or financially. It was simpler and safer just to throw away old batteries. But, surprisingly, that was not the conclusion that Energizer, Duracell, Rayovac, and other proponents of battery recycling drew in the early 2010s. Instead, they read the findings of the MIT study optimistically, focusing on the possibility that if a battery collection program was deployed nationally, it could lead to efficiencies in collection, transportation, and materials reclamation that might tip the scales in favor of recycling across all metrics. "We are trying to have some foresight," looking ten or twelve years ahead, explained Marc Boolish, an Energizer executive, in 2013. "What is the next generation of this. The ultimate one?"[124]

The industry remained concerned that if it didn't get out in front, California would be the first of many states to mandate the collection and recycling of all disposable batteries. "What we didn't want to do was to have fifty different states with fifty different regulations," a Duracell executive made clear. "That was a nightmare situation for us."[125] The industry convened a national battery summit in 2011 with the goal of creating a voluntary national collection and recycling program in partnership with retailers, municipalities, and environmentalists.[126] It also hoped to keep the process out of the legislative arena, where it feared things would quickly get "bogged down."[127]

With industry in the lead, momentum for a national recycling program built in 2011. That year, the industry leaders created the nonprofit Corporation for Battery Recycling (CBR) to oversee the recycling program. The following July, CBR issued a national call for proposals seeking a business partner to launch the program. Boolish explained, "We are seeking a stewardship organization with the capacity to build a national program that is convenient and inspires consumers to participate by recycling the batteries they use in a range of electronic and household devices."[128] The industry that had voluntarily phased out mercury was now positioning itself to launch an unprecedented, industry-funded national recycling program.

Then everything fell apart. In January 2012, Eastman Kodak filed for bankruptcy, as its core business was undercut by the rise of digital photography, and it withdrew its support. That spring, Rayovac withdrew its support for CBR, choosing instead to promote a new line of rechargeable batteries as preferable to recycling single-use batteries (even as it continued to sell disposable batteries and participate in recycling initiatives elsewhere in the world).[129] A coalition of twenty-seven environmental groups could not change Rayovac's position, despite a social media campaign, collecting twenty-five thousand letters, and a theatrical protest at its headquarters in Madison, Wisconsin.[130] Without Kodak and Rayovac participating, CBR's leaders knew it had lost the "critical mass in the marketplace" needed to "withstand the burden of free riders."[131]

That put the future of battery recycling squarely back in the legislative arena. Without major producers such as Rayovac participating and with a growing volume of imported batteries from abroad, the only viable solution was legislation mandating extended producer responsibility. That proved a hard sell. Mandatory legislation faced opposition on many fronts: toy manufacturers feared having to pay a fee every time they sold a product with batteries included; entrepreneurs like the Big Green Box recyclers worried about getting squeezed out by a policy that designated a national product stewardship organization; big players wanted the right to sue other companies to ensure industry-wide compliance; and questions remained about how and whether to legislatively set goals for collection rates and recycling efficiency.[132]

In 2014, the Corporation for Battery Recycling participated in the development of a model battery bill covering the collection and recycling of disposable batteries. Several states indicated they wanted to "fast-track" the legislation. The only state that succeeded in doing so was Vermont, which passed a law to establish a statewide collection program for disposable batteries in 2014.[133] In California, it is still illegal to throw away single-use batteries, on the grounds that they are corrosive, despite the lack of a coherent policy for managing them. In short, in 2020, most spent batteries in the United States still went into the trash. Where they were being collected for recycling, it was unlikely that was yielding much, if any, environmental benefit.

THE SHORT LIFE OF THE RECYCLED AA BATTERY

What if things had gone differently? What if the Corporation for Battery Recycling had succeeded in launching a national disposable battery recycling program in 2013? Or California had adopted extended producer responsibility legislation in 2015 and the New England states had followed suit, creating a critical mass of state-level battery programs? Those events could have provided the policy push necessary to launch large-scale recovery of batteries. But, even if that had happened, there was still a major challenge ahead: simply diverting batteries from the waste stream would do little to curb their environmental impacts.

The real challenge was what had gone largely unsaid. Despite all the talk of proper end-of-life management, how batteries were disposed of was not the problem—the overwhelming problem was how single-use batteries were sourced and manufactured in the first place. This was the clearest finding in the MIT life-cycle analysis. It estimated that production was responsible for 96 percent of the energy demand, 87 percent of the greenhouse gas emissions, 92 percent of the human health impacts, and 71 percent of the impacts on ecosystems.[134] That meant for a single-use battery recycling initiative to really make a difference, it needed to displace virgin materials used to manufacture new batteries, while using less energy and generating less pollution.

Yet, none of the proposals—whether those of the Corporation for Battery Recycling, statewide model bills, or the policies adopted in the European Union—established strict standards for what recycling meant in practice. The Vermont law, for instance, defined recycling as "any process by which discarded products, components, and by-products are transformed into new usable or marketable materials."[135] The firm contracted to manage the program sent single-use batteries to a Michigan recycling facility that recovered the zinc and manganese for use in fertilizer and steel, paper, plastic, and brass for other uses. It claimed a recycling efficiency rate of 98 percent.[136] The European Union required the recovery of at least 50 percent of the battery mass using best-available techniques. Yet, those standards still allowed recovered materials from single-use batteries to be diverted to lower-quality end uses, such as slag or road construction.

Inattention to recycling standards is hardly unique to battery policy, either in the past or today. Historically, most recycling policies and programs have been focused more on diverting recyclables from the waste stream, not ensuring that recycling maximizes, or even yields, a net environmental benefit.[137] Such lax policies contributed to the stream of recycled plastics, paper products, glass, and other materials to countries such as China, where most were downcycled—at least until 2018, when China stopped accepting such materials.

In theory, however, it was possible to close the loop on disposable batteries. Earliest discussions of such a possibility date back at least to 1918.[138] In

2015, Energizer and Retriev Technologies—which was partly owned by the same company that started the Big Green Box program—showed it could be done in practice. That year, Energizer launched the Energizer EcoAdvanced battery. What was "an impossibility for decades is now reality"—"Energizer's longest-lasting alkaline battery and the world's first AA battery made with 4% recycled batteries."[139] The company increased its 2015 advertising budget to support the launch.[140]

Although the technical details were not made public, Energizer filed for a patent in 2017 for a process to improve the recovery of high-quality electro-lytic manganese dioxide from discarded single-use batteries. The main pro-cesses detailed in the patent centered on reducing zinc and potassium contamination in the recovered manganese. Energizer released few public details about to what extent the new process actually reduced the "impact on the planet," as it claimed. But they did advertise that the new process "closes the loop," turning old batteries into new batteries.[141] Those claims were upheld by the Underwriters Laboratories and the National Advertising Division, which made specific reference to modest reductions in greenhouse gas emissions, acidification, and other environmental metrics attributable to the use of recycled materials.[142]

Then, three years after its introduction, the EcoAdvanced battery line disappeared. Energizer had invested nearly a decade of research and devel-opment in the recycled battery, starting when the European Union first mandated battery collection in 2006. It had proudly touted the invention of the EcoAdvanced as another of its technological firsts: the first flashlight, the first mercury-free battery, and, in 2015, the first disposable battery using recycled materials. Its hope was that, in the long term, recycled materials would become cost neutral or cost positive.[143] Briefly, in 2016, Energizer even announced plans to increase the use of recycled battery material to 40 per-cent by 2025.

If the push for extended producer responsibility—which Energizer had championed more enthusiastically than any of its competitors—had come to fruition in the early 2010s, Energizer's EcoAdvanced battery would have put it in the lead in closing the loop on disposable batteries. But when the efforts of the Corporation for Battery Recycling and the state battery bills largely fizzled, so too did prospects for the EcoAdvanced battery.

Energizer, of course, still sells billions of disposable batteries every year—it just makes no claims about whether they contain recycled materials. After it phased out the EcoAdvanced disposable battery, Energizer began marketing its rechargeable AA and AAA batteries as the better choice for environmentally minded consumers.[144] Indeed, that was the decision Rayovac made in 2013, when it pulled out of the Corporation for Battery Recycling in the first place. Starting in 2018, Energizer's rechargeable batteries included up to 4 percent recycled battery material, this time sourced from the types of hybrid car batteries recovered from vehicles such as the Toyota Prius.

Just as with recycling single-use batteries, however, the story on replacing single-use batteries with rechargeable batteries is complicated. The most thorough life-cycle analysis indicates that that strategy can yield a net environmental benefit across almost all environmental metrics—if the batteries have been recharged fifty times.[145] All of this is a very long-winded way of saying that if you are still unsure about what to do with your disposable batteries or whether you should replace them with rechargeable AA or AAA batteries, that is understandable. There are no good choices.

In researching the early history of disposable batteries, I came across numerous articles that catered to the tinkerer, offering step-by-step instructions for how to make a zinc-carbon dry-cell battery. So, to bring my research on the history of disposable batteries to a close, I decided to make one myself. A 1913 *Scientific American* article titled "Dry Batteries and How to Make Them" outlined the materials needed: zinc sheet for the anode, "black oxide of manganese" for the cathode, "chloride of ammonium" for the electrolyte, sawdust, ground carbon, and water. It recommended buying materials at an electrical supply store; I purchased them all online from a laboratory supply company. It also recommended using a solution of "nitrate of mercury," but I decided to skip that step, since I didn't need a battery with a shelf life. The basic procedure was to roll the zinc sheet into a cylinder, line the inside of the cylinder with paper, and pack the manganese dioxide–carbon mixture, wetted with the ammonium chloride electrolyte, around a carbon rod inserted in the center of the zinc cylinder.[146]

If successful, my makeshift battery would power a small flashlight bulb. A current of electrons would flow from the zinc sheet, through the bulb, to the electron-hungry manganese dioxide. Whether in the 1910s or today, no matter whether I assembled a dry-cell the size of a AAA battery or tried to create one of the large-format No. 6 batteries, the underlying electrochemistry of the zinc and manganese dioxide yielded a battery with an electric potential of roughly 1.5 volts. That was a product of the basic electrochemical properties of the materials. The *Scientific American* article obscured all of the trial and error that had gone into the creation of early dry-cell batteries.

Much harder than assembling the battery was tracing the source of the materials that made my desktop experiment possible. In 1913, it would have been possible, with some degree of certainty, to trace the zinc, manganese dioxide, and other materials to their source. At the time, most zinc was mined in the Tri-State District at the borders of Oklahoma, Missouri, and Kansas, and most battery-grade manganese dioxide came from the Caucasus Mountain region in Central Asia.

Compared to the early nineteenth century, however, today's materials are as much made as they are mined. This had always been true to an extent— ores always required some processing—but the rise of electrolytic refining helped transform the metals industry. This is important, because even as miners rely upon ever-lower-grade bodies of ore, many industries require purer materials. Once mining depended upon locating the highest quality seams of ore. Now where the ore is sourced matters less, and how the ore is processed and refined matters more. It was an early lesson from the manganese mines in Montana, which could only replace the higher-quality ores from the Caucasus region as a result of better processing.

There are two broad consequences of this transition. First, as processing became more important than mining, it became harder to trace the materials back to any one place. Indeed, no matter how many calls I made, I could learn almost nothing about the origins of the zinc, manganese, or other materials from which I assembled my homemade battery. Second, as processing became more important, so too did the amount of energy needed to run these processes. In short, the purer the materials required, the higher the level of processing, the more energy intensive and environmentally costly the materials became. This is what I call the paradox of purity.

The increased importance of processing points to the need to reconceptualize what we think of as a "mine." On the one hand, as lower-grade ores become suitable raw materials for processing, the scope of the mine gets larger and larger.[147] But why mine low-grade "natural" ores, which may run less than 1 percent zinc or manganese, when we could mine the waste stream, where the concentration of metals in old batteries, cell phones, computers, and other products is already an order of a magnitude higher than what is found in nature? On the one hand, Energizer's success in recovering high-quality electrolytic manganese dioxide—a feat many people thought impossible—from spent batteries makes clear the prospects for urban mining. But it is also important to remember that recycling those materials yields uncertain benefits relative to conventional mining.

This is a clear reminder that recycling is not a panacea for the materials challenges we face. Just as it takes significant investments of energy to refine raw materials, it also takes significant investments of energy and other resources to recover recycled materials. In the case of lead-acid batteries, recycling is clearly preferable to disposal. But that is not clear in the case of single-use AA and AAA batteries. The other surprising lesson of efforts to recycle disposable batteries is how much the seemingly trivial issues of collecting, sorting, and transporting disposable batteries actually matter. Right now, the present inefficiencies in how disposable batteries are collected likely outweigh the benefits of recycling. For recycling to succeed, it needs to be approached not just as a narrow technical challenge to be solved by chemists and engineers but as a social challenge, one that requires improvements in the efficiencies in every aspect of the system, from how batteries are collected in homes and aggregated at recycling depots to how they are transported to waste processors. Only then will it be clear that recycling disposable batteries is, in fact, better than just pitching them in the trash.

3

LITHIUM-ION BATTERIES, THE SMARTPHONE, AND A WIRELESS REVOLUTION

IN THE FALL OF 1958, ARNE LARSSON'S HEART WAS FAILING. AT TIMES, his heart beat once every three seconds, leaving him weak and prone to fainting spells. The Swedish engineer was only forty-three years old, but a viral infection had worsened a congenital heart condition. His doctors tried a range of treatments, including ephedrine, atropine, and caffeine. On bad days, his wife and aides had to resuscitate him up to thirty times. The weak spells could last for weeks, and doctors feared for his life.

On October 8, 1958, Swedish doctors implanted the first internal pacemaker in Larsson's body. The device—made possible by transistors and advances in battery technology—was designed to pace Larsson's heart at sixty-five beats per minute and restore his health. But Larsson's first pacemaker failed within three hours. A replacement worked for a week before it also failed. Maintaining Larsson's pacemakers required eighteen surgical interventions between 1958 and 1978. The most common source of Larsson's pacemaker failures? The batteries.[1]

That changed in 1978, when Larsson received his first lithium-powered pacemaker. The new lithium battery replaced the mercury zinc batteries used in earlier pacemakers. It had double the energy density, negligible self-discharge, and could be hermetically sealed, which made it much more reliable. Engineers projected that a lithium battery could power a pacemaker for 340 million heartbeats—more than a decade—without replacement.

That combination of reliability and longevity expanded the possibilities for battery-powered medical implants. After the mid-1970s, lithium-based batteries became the standard for pacemakers, defibrillators, and other implants that have helped millions of patients to live fuller and healthier lives.[2] Although Larsson underwent surgery five more times before he died in 2001 of cancer, none of those interventions was to replace a dying battery.

The story of the pacemaker illustrates the potential of high-power, long-lived lithium-based batteries. As early as the mid-1960s, boosters began championing the possibility that future lithium batteries could one day power electric cars and high-energy storage batteries. Although battery chemists succeeded in developing the small, nonrechargeable lithium batteries used in pacemakers by the late 1970s, developing larger-format rechargeable lithium batteries capable of powering portable telephones, laptop computers, or electric cars proved a bigger challenge.

What drew electrochemists to lithium was the combination of light weight and high performance. With only three protons, four neutrons, and an atomic weight of seven, lithium is the third-lightest element after hydrogen and helium, making it the lightest of all the metals. Of its three electrons, the outermost electron is only weakly bound to the nucleus, which makes lithium a willing participant in the kinds of electrochemical reactions that charge and discharge a battery. Of all the elements, lithium packs the most electrochemical punch per unit weight.

But harnessing lithium's desirable electrochemical properties has meant managing its least desirable property, which is the risk of fire. If a lithium-ion battery short-circuits or overheats, there is a risk the battery's organic electrolyte can catch fire. As one former Tesla engineer explained, when a lithium-ion battery goes, "it is a scary sight," no matter how many times you have seen it.[3] Sometimes a malfunctioning lithium-ion battery releases a curl of smoke and swells before it catches flame. Other times, the battery just explodes.

Despite these risks, which have been made manageable by careful engineering, rechargeable lithium-ion batteries have been deployed by the billions, providing a safe, reliable, and convenient source of portable energy since Sony first introduced them to the market in 1991. In telling the story of the surprisingly rapid rise of the rechargeable lithium-ion battery, this chapter focuses on two points.

First, public attention on batteries often centers on the potential for breakthroughs: the newly invented battery that can be made of rhubarb, charged in minutes, or store an order of magnitude more energy. Yet, the story of lithium-ion batteries is not about a "breakthrough" technology. Instead, it is a story of incremental improvements in performance, safety, and cost that made the lithium-ion battery just "good enough." That a technology that is "good enough" could be so rapidly adopted is explained by the lithium-ion battery's role as an enabling technology. It played a key role in enabling a new generation of portable devices that were faster, longer-lasting, and more useful. Incremental gains in battery performance were multiplied by the extraordinary pace of innovation in consumer electronics between the 1980s and early 2000s. Together, these advances enabled a new culture of mobility that revolved around portable devices that were always on and always connected.

Second, bringing lithium-ion batteries to the global market at scale reflected a shift in the global economy and the organization of resource extraction and manufacturing in the late twentieth and early twenty-first centuries. It depended upon resources from developing countries, such Chile and the Democratic Republic of the Congo. It depended upon increasingly nimble manufacturing processes centered in Asia that could be scaled quickly and cheaply. And it depended upon global policies and institutions that promoted privatization, global investment, and free trade, which allowed companies like Motorola or Apple to outsource their manufacturing with minimal concerns for workers or the environment. In short, the history of the lithium-ion battery has been about more than just portable power; it has been driven by a profound shift in global economic and political power that put the United States in the position of playing catch-up to China, Japan, and other Asian countries in the race to manufacture advanced batteries and secure a toehold in a clean energy future.

A LONG-OVERDUE NOBEL PRIZE

The Swedish Academy of Sciences awarded the 2019 Nobel Prize in Chemistry to three scientists for their pioneering roles in researching the basic science underlying the modern lithium-ion battery: Stanley Whittingham and John Goodenough of the United States and Akira Yoshino of Japan. The

Nobel Prize recognizes researchers who have "conferred the greatest benefit to humankind."

The research the Nobel Prize Committee recognized had been conducted in the 1970s and 1980s in the United States, England, and Japan. The impetus for that early research had not been concerns about global warming but, rather, the oil shocks of the 1970s, which drove up the price of oil, left Americans lined up at gas stations to fill up their cars, and raised concerns about oil shortages. At the time, existing battery technologies, including lead-acid batteries and nickel-cadmium batteries, lacked the power, energy capacity, and reliability needed to power a viable alternative to conventional cars.

It was for that reason that Exxon opened up a new research laboratory in the early 1970s and recruited Stanley Whittingham, a thirty-two-year-old postdoctoral researcher from Stanford University, to run a new Advanced Battery Project in 1973. The lab was located in Linden, New Jersey, in the heart of Exxon's hellish industrial sprawl of refineries and storage tanks. Whittingham recalls that Exxon invested in research at the time just "like they invested in drilling oil."[4] They expected only one in five projects to pay off. Whittingham's research, which centered around solid-state chemistry, showed early promise.

In 1976, Whittingham published a paper in *Science* that outlined the basic properties of a reversible, energy-efficient, rechargeable lithium battery made of abundant materials that operated at room temperature.[5] That was an important advance, since similar experimental batteries only functioned at more than 500°F. The research was the culmination of a burst of activity at the Exxon labs centered around the potential for pairing a titanium-disulfide cathode with a lithium metal anode in a battery. The secret to titanium disulfide was its molecular structure. A lattice structure allowed lithium ions to be stored in between the layers of titanium disulfide and then released at a rapid rate without degrading its structure.

Exxon immediately began eyeing the potential application of titanium disulfide in a durable, high-energy battery system to power an electric car. It poured millions of dollars into the Advanced Battery Project in the 1970s. Considering that Exxon remains one of the world's largest petroleum companies today and is known for its support of global-warming skeptics, this history may be surprising.[6] But in the 1960s and 1970s, the corporate culture at Exxon was different. The company pioneered research on climate change and invested in alternative energy technologies. In 1977, Exxon exhibited its rechargeable lithium battery at the nation's first electric vehicle exhibition in

Chicago.[7] In the late 1970s, Exxon entered into a joint technology agreement to develop a gas-electric car with Toyota (a partnership that may have helped plant the early seed for the Toyota Prius, which debuted sixteen years later).[8]

But Exxon's enthusiasm for these projects disappeared when oil prices stabilized in the 1980s. Whittingham's early research never resulted in a commercially viable battery. Benchtop experiments involving the lithium metal anodes kept catching fire, and the substitutes Whittingham's group tried all undermined the battery's longevity or performance. Despite the setbacks, Whittingham had demonstrated the potential of what he described as "intercalation" chemistry—the ability of a crystalline lattice to store energy by accepting and then releasing ions with minimal structural change.

What makes intercalation chemistry so important is that the electrode's lattice structure functions as a figurative bookshelf for lithium ions—checking lithium ions out as the battery charges, and then allowing them to be checked back in when discharged. In this analogy, the best bookshelves were light, sturdy, and durable; could pack in the most lithium ions possible; and allowed for rapid circulation.[9] But much work remained to identify the specific materials that would result in a safe and reliable lithium-ion battery.

If you have ever heard John Goodenough interviewed, one thing that likely stood out is his laugh: it is the cackle of a most genial scientist. Goodenough, a physicist by training, knew about Whittingham's research when he took up a professorship in chemistry at Oxford University in 1976. Since then, Goodenough has made his career as a solid-state materials scientist at the interface of physics and chemistry, researching semiconductors, metal oxides, and, since the 1970s, battery electrodes. At the time, Goodenough believed it was "obvious" that he should be working on energy storage. That conviction had not ebbed when he was awarded the Nobel Prize at age ninety-seven, making him the oldest laureate in history.

In Goodenough's view, solving the problem with Whittingham's fire-prone lithium batteries required changing the source of lithium in the battery, eliminating the volatile lithium metal anode, and using lithium compounds in the cathode instead. Between 1976 and 1987, Goodenough, working with three different visiting scientists at his laboratory, discovered three distinct classes of potential lithium-ion cathodes, each intercalation materials that paired lithium ions with metal oxides of cobalt, manganese, or iron, respectively. These discoveries provided the foundation for the lithium-ion battery industry. Each cathode would be commercialized in the decades to come.

The Goodenough lab's breakthrough made it possible to eliminate metallic lithium, improving the safety and stability of lithium-ion batteries. Using a metal oxide, instead of a sulfide, also doubled the potential voltage of a lithium-ion battery. The most immediate downside was that lithium compounds came with additional molecular baggage, in the form of metals such as cobalt, manganese, and iron, which reduced their energy density. In the case of the lithium–cobalt oxide, for instance, no more than half of the lithium ions could be checked out during charging without damaging the cathode. Even with those limitations, however, the lithium-ion batteries had the potential to store twice as much energy per unit weight and operate at double the voltage of existing battery systems.

Unlike other battery chemistries, such as lead-acid or disposable alkaline batteries, which rely on a very specific combination of anode, cathode, and electrolyte, Goodenough's work laid the groundwork for a variety of different lithium-ion battery chemistries. Indeed, by the early 2000s, at least six lithium-ion battery chemistries had entered into commercial production, most of which were variations on the Goodenough laboratory's research. Each deployed a cathode tailored to deliver a different balance of properties, such as durability, power, energy density, safety, and cost. Of the different chemistries, the lithium–cobalt oxide battery became the workhorse of the wireless revolution in the early 1990s.

Developing a workable lithium-based battery, however, also required a safe and stable anode to pair with the Goodenough lab's cathodes. In the early 1980s, researchers discovered that layered graphite allowed for the intercalation of lithium at the anode. The advantage of graphite was that it had what electrochemists describe as a high redox energy, which could be paired with the low redox energy of a lithium–cobalt oxide cathode. To return to the bookshelf analogy, it was as if the graphite-anode bookshelf was located on the fourth floor of the library, while the lithium–cobalt oxide cathode bookshelf was located on the ground floor. That difference in redox potentials is what gave the lithium-ion battery its high voltage.

But graphite also degraded when paired with the most commonly used electrolyte, propylene carbonate. That resulted in experimental batteries with short lives. The solution to that problem came from Akira Yoshino, a Japanese engineer working for Asahi Kasei, one of Japan's largest petrochemical companies, which specialized in synthesizing plastics, fibers, and, soon, battery electrodes. Yoshino, who was trained as a petrochemical engineer, came to battery research with a focus on carbon-based electrodes. Yoshino recounts being frustrated with his research until he stumbled across Goodenough's paper on lithium–cobalt oxide cathodes in the early 1980s. In retrospect, the timing was fortuitous. As he later recalled, lithium–cobalt oxide was "just what I wanted." And Yoshino began pairing lithium–cobalt oxide with a range of what he categorized as "carbonaceous materials."[10]

By 1987, Yoshino had working prototypes of a lithium-ion battery that paired a lithium–cobalt oxide with a high-purity petroleum coke, or synthetic graphite, baked to just the right level of density and crystallization. The resulting anode was sufficiently layered to intercalate lithium, but the disordered structure included irregularities where the carbon bound together with strong covalent bonds, providing the necessary structural strength to cycle repeatedly. Or, to return to our analogy, it was a functional and durable fourth-floor bookshelf. Asahi Kasei sent Yoshino to the United States to produce the first batch of lithium-ion batteries, far away from competitors such as Sony or Panasonic. The prototypes withstood basic safety tests, maintained a high voltage, and proved capable of being charged and discharged repeatedly. In Yoshino's view, this was "the moment when the lithium-ion battery was born."[11]

Bringing rechargeable lithium-based batteries to market happened surprisingly quickly, but not without missteps. A British Columbia–based company, Molicorp, moved aggressively to produce a lithium-metal battery, similar to what Whittingham experimented with at Exxon in the late 1980s. Despite a decade of engineering to address the safety risks, Molicorp drew headlines when a cell phone using the newly released Molicel caught fire. That forced a recall, ended production of the new battery, and scared off companies such as Sanyo and Panasonic, which had been pursuing similar strategies. Despite their potential advantage, many companies viewed lithium-based batteries as too risky.[12]

Japan-based Sony Corporation became the driving force in bringing a lithium-ion battery to market in the late 1980s. Since the late 1970s, with the breakout success of the Sony Walkman portable music player, growing concerns over battery-related mercury pollution in Japan, and the expanding market for portable electronic devices, Sony had taken a keen interest in developing a high-power rechargeable battery. When Sony launched an ambitious effort to develop a "dream battery" in 1987, it got a head start from Asahi Kasei.[13] In January 1987, Sony-Energytec signed a nondisclosure agreement with Asahi Kasei, gaining exclusive access to Yoshino's working prototype. That became the basis for four years of intensive engineering work at Sony to mass-produce a safe and reliable lithium–cobalt oxide battery.[14]

When Sony released the first lithium-ion battery in June 1991, it put Sony in the lead of what the *Wall Street Journal* described as the race "to develop new technology for long-lived, lightweight, environmentally friendly rechargeable batteries" that could transform everything from "consumer electronics to electric cars."[15] At the time, many observers remained skeptical of the safety of lithium-based batteries, and other companies, such as Sanyo and Panasonic, invested research and manufacturing capacity in an alternative nickel–metal hydride battery (NiMH) chemistry, which was safer but stored only half the energy of a lithium-ion battery. By the time the Nobel Prize Committee awarded Whittingham, Goodenough, and Yoshino the Nobel Prize in 2019, it was clear that lithium-ion batteries had not just out-paced nickel-cadmium and nickel–metal hydride batteries; they had lapped them. In the words of the Nobel Prize Committee, their research had laid the groundwork for a "phenomenal battery" that enabled the "wireless revolution" of the early twenty-first century.[16]

THE WIRELESS REVOLUTION

It is easy to appreciate the Nobel Committee's sense of wonder for the transformative role of lithium-ion batteries. In 1991, when Sony put the first lithium-ion batteries onto the market, cell phones were still so bulky they were generally found in cars, not in pockets. That year, Apple released the PowerBook 100, an early mainstream laptop computer. It weighed five pounds, measured two inches thick, and ran for two to four hours with a fully charged lead-acid battery.[17] And the internet was still largely the province of researchers and primarily used for specialty applications in higher education, research, and government. Few people anticipated that it would become the digital fabric of daily life, stitching together devices and the people who used them worldwide.

As the Nobel Committee noted, for all the importance of the lithium-ion battery, it only played an enabling role in the wireless revolution. Exponential improvements in digital computing performance drove much of this revolution. In 1965, Gordon Moore, a cofounder of Intel, predicted that the number of transistors that could be packed onto a silicon chip would double every two years as a result of improvements in design and manufacturing that would, in turn, drive down costs. Those advances in processing power enabled a culture of mobility in which access to information became increasingly ubiquitous and instantaneous.[18]

To put that transition in context, consider this comparison: A smartphone in 2020 delivered more processing power than did the world's fastest supercomputer in 1991. What is less often noted, but just as important, is that a smartphone delivered that processing power at 0.002 percent the cost, 0.003 percent the weight, and using 0.001 percent as much energy as the 1991 supercomputer. Now, consider this back-of-the-envelope estimate. If the best-selling automobile in 1991, the Honda Accord, had improved at the same rate in 2020, the car would have cost less than a dollar, had a top speed of 1,200 miles per hour, and traveled twenty-three million miles on a gallon of gasoline.[19]

It is that last data point that is so telling. The extraordinary advances in the energy efficiency of microprocessors help explain why incremental improvements in battery technology were so consequential. Those gains were, in effect, being multiplied by the gains in the digital technologies they powered. The 18650 battery cell offers a useful metric for tracking the improvements in lithium-ion battery technology. In battery nomenclature, 18650 means the battery cell is 18 millimeters in diameter and 65 millimeters in length, and the 0 signifies that it is a cylindrical battery. That makes it just slightly larger than a single-use AA—just large enough that you cannot accidentally insert a lithium 18650 cell into your remote control.

Between 1995 and 2010, the energy capacity of an 18650 cell tripled as a result of tweaks to battery chemistry, engineering advancements including thinner separators and current collectors, and coatings on the anode and cathode that improved stability and durability. Over the same time period, the price of lithium-ion batteries fell by 80 percent, as manufacturing expanded and manufacturers realized economies of scale. The 18650 cell formed the basic unit of most lithium-ion battery packs, which could be packaged together to power a camcorder or early laptop in the mid-1990s or, more recently, a power tool or an electric car.[20]

One other advance contributed to the rapid uptake of lithium-ion batteries and the advent of the wireless revolution. In 1999, the Swedish company Ericsson introduced a cell phone powered by a lithium-polymer battery. Unlike the bulkier 18650 cylindrical battery cells, lithium-polymer cells were flat, measuring three to four millimeters thick, and could be manufactured in a variety of shapes.[21] That made it possible to design the battery to fit a device, rather than designing the device to accommodate the battery. That advance launched a trend toward ever-thinner portable devices that accelerated with the advent of smartphones, such as the iPhone in 2007, and proliferated with tablet computers, thin laptops, and other portable electronics in the 2010s.

In short, to fully understand what made lithium-ion batteries so "phenomenal," it is helpful to return to an idea that I introduced at the start of this book: batteries matter not just for the quantity of energy they deliver but also for other qualities, including reliability, rechargeability, toxicity, and cost. And in every category, lithium-ion batteries represented an improvement over existing rechargeable batteries. Lithium-ion batteries proved more reliable than earlier battery technologies, capable of being charged twice as many times. They did not exhibit a "memory effect," meaning they could be charged and discharged when most convenient. And they were less toxic. Although common lithium-ion battery chemistries employed cobalt and nickel, which posed some health concerns, they did not rely on more toxic materials, such as lead, cadmium, and mercury.

Their primary downside was the price. Yet, by the early 2000s, lithium-ion batteries dominated the surging demand for rechargeable batteries for mobile applications, despite their high cost. In 2000, lithium-ion cells still cost double a nickel-cadmium or nickel–metal hydride battery, even factoring in their higher voltage and energy density.[22] Materials accounted for roughly two-thirds of manufacturing expenses. But by 2006, as the scale of production grew, the price of lithium-ion batteries reached parity with existing battery technologies. And by 2010, lithium-ion batteries accounted for almost 80 percent of a vastly expanded market for rechargeable batteries, providing the portable power that allowed people to text, snap digital pictures, surf the web, and, of course, make a telephone call.[23]

TAKING APART A LITHIUM-ION BATTERY

To better understand what goes into a lithium-ion battery, I took one apart. I started with an Apple MacBook computer that I bought in 2007 that had been collecting dust on my bookshelf. Unlike recent Apple products, the battery pack was a user-replaceable part, measuring half an inch thick, three inches wide, and eight inches long. The label indicated it was a 10.8-volt mercury-free Li-ion battery pack.

On the outside, the battery pack was a sleek package form-fitted to slot into the bottom of the computer. It looked different on the inside. Six

Chinese-manufactured lithium-ion polymer battery cells were packed inside a protective aluminum shell. Individual cells are the building blocks of a lithium-ion battery pack, whether for a laptop, a power tool, or an electric car. Wiring batteries in series—meaning connecting the positive to negative terminals of sequential batteries—increases the voltage of a battery pack. In the laptop, three cells were wired in series to form a battery string delivering 11.1 volts, just over the pack's promised 10.8 volts. Wiring battery strings in parallel—meaning connecting the like terminals—increases energy capacity. In the laptop, two strings of three cells each were wired in parallel to provide 55 watt-hours of capacity.

After using a voltmeter to confirm that the battery was dead, I disassembled one of the six foil-wrapped battery cells. Each cell was made up of a three-foot-long, two-and-a-quarter-inch-wide strip of layered film, which had been folded sixteen times into a package about two inches on a side and a quarter-inch thick. This was the electrochemical heart of the lithium-ion battery. The film was made up of three layers: the anode, the cathode, and the separator. Both the anode and cathode looked identical: long strips of tissue paper–thin metal foil coated with thin layers of electrode material—in this case, likely a lithium–cobalt oxide cathode and graphite anode. The entire package smelled of chemicals, since it was still moist with electrolyte applied when it was manufactured, thirteen years before.

My old MacBook battery offered few clues as to its origins. Although it was labeled a mercury-free Li-ion battery, it included no specific information regarding its material composition or where those materials had been sourced from. What it did yield were clues about how the global economy was put together in the early 2000s. The battery's label read: "Designed by Apple in California. Assembled in China." Since the 1980s, a trend toward outsourcing manufacturing allowed big-brand consumer electronics companies to avoid investing capital in manufacturing capacity by relying on specialized manufacturers that could realize greater economies of scale, more rapidly ramp production up and down to match demand, and locate manufacturing plants in regions with lower costs, weaker regulatory oversight, and abundant cheap labor. When Apple's iPhone boomed in the early 2010s, it drew substantial public attention to working conditions at Foxconn City, the sprawling manufacturing plant and company town owned by Taiwan-based Foxconn in Shenzhen, China, that employed 230,000 workers.[24]

What drew far less attention, however, were the extensive networks of suppliers that mined and processed the minerals, materials, and other components upon which China's low-cost consumer electronics industry depended. My MacBook battery may have been assembled in China, but to understand what made it possible means tracing its origins to growing supplies of specialized materials, including lithium, cobalt, and graphite, mined in countries such as Chile, Australia, the Democratic Republic of the Congo, and China itself. In the same ways that a global economic order increasingly promoted free trade and foreign direct investment with minimal safeguards for workers or the environment in the 1990s, the same set of rationales also underpinned the intensive development of extractive industries that supplied the metals, chemicals, and other materials important to the lithium-ion battery industry. In this case, the batteries were included.

LITHIUM

The cosmological origins of lithium have long been a mystery. Until 2020, the general theory was that the vast majority of the lithium originated in the minutes following the Big Bang, along with helium and hydrogen, making it one of the lightweight elements whose origins predated the formation of stars and galaxies. But astrophysicists had long been stumped. Observations revealed far more lithium in the galaxy than theory predicted.

"Given the importance of lithium to common uses like heat-resistant glass and ceramics, lithium batteries and lithium-ion batteries, and mood altering chemicals, it is nice to know where this element comes from," explained Sumner Starrfield, the astrophysicist who led the team that resolved what was known as the "lithium problem."[25] Based on their analysis, published in 2020, the vast majority of lithium comes not from the Big Bang but from the same place as most other elements—the stars.

Starrfield's team confirmed that lithium-7—the most common lithium isotope—is the by-product of common stellar explosions, known as classical novae. These fusion-powered explosions originated when a dwarf star drew off hydrogen from a companion star, compressing it until a fusion-powered reaction began. That reaction generated a pulse of energy that ejects newly formed elements, including lithium-7, into space.[26]

Lithium is the twenty-fifth most abundant element on Earth. Since pure lithium reacts violently with water, naturally occurring lithium is found in minerals or salts. Although lithium is spread widely across the planet, economically recoverable deposits of lithium are generally found in two forms:

in granite rocks known as pegmatites formed some 2.5 to 3 billion years ago and in salt flats formed in the past million years by geothermal activity and runoff.[27]

Until the mid-1990s, the United States led the world in both the mining of lithium and its consumption—a legacy of the Cold War nuclear arms race. Enriched lithium fueled thermonuclear weapons then and today. In fact, when physicists underestimated the volatility of lithium isotopes in the first lithium-fueled thermonuclear weapon, they unleashed a test explosion that was nearly three times as powerful as predicted. That bomb, tested in the Bikini Islands in 1954, produced more power in a single detonation than all the bombs Allied forces dropped in World War II.[28]

Lithium also played a small role in the American household in the postwar era, largely for its thermal and medicinal properties. It was used in specialty glassware products, including oven-safe bakeware and television tubes. It lowered the energy demand during aluminum smelting. And it was valued for its therapeutic properties: it was increasingly found in the medicine cabinet of those suffering from manic-depressive illness, now known as bipolar disorder, after the 1970s.[29]

Through the late 1980s, most of the world's lithium was supplied from hardrock mines in North Carolina and salt brines in Nevada. But the geography and scale of the lithium industry began to change rapidly in the 1990s, as the industry anticipated growing demand for new lithium-powered batteries. Boosters set their sights on the world's largest-known reserves of lithium in South America's "Lithium Triangle."

Since the 1970s, geologists knew that the vast high-elevation salt pans that lined the volcanic mountain chains along the north-south-trending borders of Argentina, Bolivia, and Chile contained high levels of lithium and other minerals.[30] These vast salars, many of which measure more than one thousand square miles in size, are the remnants of ancient seas. Their hardened saline surfaces, measuring a few inches to a few feet thick, were formed over millennia by evaporation and the deposition of salts. What miners coveted lay just beneath the surface. Subsurface brines made up the largest-known reserves of lithium in the world.

In the early 2000s, the region became known as the Lithium Triangle. But that description obscured a longer history of mining. A century before, the region had been known as the "Nitrate Desert" because of its abundant stores of sodium nitrate, more commonly known as Chilean saltpeter, which

Europeans increasingly valued for the manufacture of fertilizer and gunpowder. British investors started financing numerous saltpeter mines in the mid-nineteenth century, Chile went to war with Peru and Bolivia to secure its claim to the region in the 1880s, and Chileans took over the saltpeter mines at the start of the twentieth century. The saltpeter industry drove the country's economic growth and made Chile the world's leading source of sodium nitrate.[31]

Chile's saltpeter industry collapsed after World War I. The availability of synthetic fertilizers (produced through an energy-intensive process fueled by natural gas) drove down prices, destabilized Chile's resource-dependent economy, and left behind ghost towns in the Atacama Desert. By the 1940s, copper had replaced saltpeter as Chile's most important mineral export (as it still is today). But it was out of the remnants of the saltpeter industry that Chile's lithium industry emerged. Between the 1930s and 1970, Chile consolidated the saltpeter industry under the Sociedad Química y Minera de Chile (SQM).

SQM began to develop the Salar de Atacama, Chile's largest salt flat, in the 1990s to secure its own source of potassium chloride—an essential raw material for converting saltpeter into nitrate fertilizers.[32] Instead of mining rock, however, SQM mined the salar for its mineral-rich brine. It pumped subsurface brines into plastic-lined solar evaporation ponds. Between 1992 and 1996, SQM constructed a sprawling industrial operation that spanned more than eight square miles—an expanse six times larger than New York's Central Park—making it one of the world's largest producers of potassium chloride. Yet, the brine SQM pumped from beneath the Salar de Atacama contained more than just potassium chloride. It also contained a small fraction, 0.2 to 0.4 percent, lithium.[33]

In 1996, anticipating growing demand for lithium-ion batteries, SQM bet big on its ability to produce lithium as a by-product of its potassium chloride operations. The company promised to double global lithium production and drive down prices. SQM's optimism was driven by the favorable conditions in the Atacama Desert. Early estimates indicated the Salar de Atacama sat atop millions of tons of lithium. And, equally important, the region was literally the sunniest place on Earth. Annual rainfall measured less than half an inch per year.[34] That allowed SQM to process brine all year-round. Unlike most other mining operations, which relied on fossil fuels to concentrate minerals, what SQM needed most to concentrate brine for processing was ample space for the evaporation ponds and time.

FIGURE 3.1 The scale of lithium evaporation on the Salar de Atacama, Chile, is enormous. These silhouettes illustrate the footprint of the Tesla Nevada Gigafactory, planned to be the largest factory by footprint in the United States (A), compared to the size of New York's Central Park (B) and SQM's evaporation pond complex on the Salar de Atacama in Chile (C).

The political winds also blew in SQM's favor on the Chilean desert. Starting in the 1970s, with the rise of the military dictator Augusto Pinochet, Chile embraced free-market policies that encouraged foreign investment, privatization, and a resource-based export economy. That made Chile a testbed for neoliberal development policies that increasingly linked developing countries to global markets. Although a 1983 Chilean law declared lithium a strategic mineral of national interest, SQM was one of two companies—the other being US-based Foote Minerals—that acquired mineral rights to the Salar de Atacama that predated the law. Those rights came with favorable terms, including low royalties and taxes, meant to incentivize private investment that last through 2030.[35]

Yet, turning the Salar de Atacama's brine into lithium carbonate—the key raw material for battery cathodes and other applications—was a finicky exercise in applied chemistry. It hinged on mixing brines from different wells to improve the recovery of lithium. It required removing salts produced during the evaporation process, including valuable potassium chloride. And it required additional chemical processing, including the addition of lime, which precipitated out impurities such as magnesium. It took approximately a year of solar evaporation to distill the brine into a concentrated liquor that was 6 percent lithium. Just as with the electrolytic manganese dioxide important to disposable AA batteries (see chapter 2), lithium carbonate was as much manufactured as mined.

To process that liquor into a marketable raw material, SQM located its lithium plant in the Salar de El Carmen, near the port city of Antofagasta, Chile. There it built additional evaporation ponds that concentrated the brines left over from potassium chloride production in the Salar de Atacama into a lithium-sulfate rich solution, which had a lime-green hue. The final step was heating the brine with soda ash, a form of sodium carbonate sourced from Wyoming's trona fields, which yielded a battery-grade lithium carbonate, the precursor for manufacturing a lithium-ion battery cathode. In 1996, SQM produced its first batch of lithium carbonate—8,000 metric tons. By 2002, it had scaled production to 21,500 tons, meeting nearly 40 percent of global demand.[36]

SQM's rapid success hinged on its ability to produce lithium at half the cost of other firms. Most analyses rightly emphasized the efficiency of brine-based solar extraction processes. Brine-based lithium extraction required less energy and inputs than lithium sourced through traditional hardrock mining (which would become increasingly important to the global lithium supply in the 2010s).[37] SQM emphasized the "sustainability" of its solar-based operations. It claimed that more than 90 percent of the energy it consumed came from the sun.[38] What drew less attention, however, were the social consequences of the water-intensive brine-based water extraction processes on the Salar de Atacama.

Historically, the Salar de Atacama had been sparsely populated by indigenous peoples who used the land for grazing, farming, and salt collection. SQM's operations did not so much displace indigenous peoples as reorient nearby communities toward a market-based economy. Indigenous communities clustered at the edges of the salars, where mountain-fed streams delivered a precious and scarce flow of water. As the anthropologist Cristian Herrera documents in an ethnography of the Atacameño community of Toconao, lithium-mining activities on the salt flats eroded the community's economy and culture—employing workers, supporting local entrepreneurship under the guise of corporate stewardship, and acquiring water rights—even though Toconao was located fifty miles distant from SQM's active mining operations.

SQM's impacts on neighboring communities stemmed from Chilean state policy, which sanctioned privatization of mining concessions and water rights favorable to extractive industries. Although the Indigenous Law of 1993 allowed for Atacama communities to claim and manage water rights as communal property, which Toconao did, the state allocated the vast majority of land and water rights in the region to mining companies. And that trend toward privatization had ripple effects. The communal farming and irrigation association just south of Toconao entered into a long-term lease to supply its limited water resources to SQM. As local agriculture practices became less important, farmers began to hire out their ritual responsibilities for maintaining Toconao's communal irrigation infrastructure.[39] And many in the community worried that as SQM used more water, it would disrupt the region's water cycle, leading to shortages and hardship, especially in a community that increasingly treated water as a commodity rather than a shared resource.

As SQM drew water and workers away from the Atacameño communities, the company emphasized the support it offered in return. By 2008, in Toconao, only half of the residents still farmed and tended orchards. Many people instead worked jobs directly or indirectly tied to the mining industry and its community initiatives. Through its corporate social-responsibility outreach programs, which SQM touted in glossy annual reports written in

English and aimed at an international audience, SQM highlighted its out-reach activities: Microloans encouraged local entrepreneurship and supported the development of a small-scale local wine industry. SQM championed its commitment to being a "good neighbor," emphasizing its commitment to hiring local workers and providing community support, including funds for local museums, numerous educational programs, and scholarships for professional development. It also supported sustainable agricultural initiatives, including the "Atacama Fertile Land Program," which promoted water-conserving drip-irrigation systems.[40] Such programs were important contributions to these communities, although small in scale for a company that averaged more than $2 billion in annual revenue in the 2010s.

Most uncertain about the brine-based lithium operations were the environmental impacts. The brine-based operations were both land and water intensive. The footprint of SQM's solar ponds in the Atacama Desert had grown to nearly forty square miles by 2010, thirty times the size of Central Park, and the volume of groundwater it used was approximately one-fortieth of New York City's annual water usage. Although SQM had extensive operations monitoring air emissions, water consumption, and other impacts, uncertainty centered around the potential long-term impacts on local biodiversity and the region's hydrology. Species that relied upon the salar included a variety of flamingos, lizards, and the endemic tamarugo tree.[41] The tamarugo, a spindly legume, is the only tree that manages to eke out a living on the dry salt flats. It taps into the groundwater below.[42] Researchers warned that groundwater pumping would create a cascade of ecological effects, jeopardizing the tamarugo trees, the biological productivity of the salar's few lagoons, and the habitat of well-known species such as the flamingos.[43]

By the early 2000s, SQM's operation was one of three brine-based operations in the Lithium Triangle, which altogether accounted for about half of global lithium production, and a larger percentage of the lithium carbonate, most important for battery production. Despite the outsized importance of lithium to lithium-ion batteries, however, it is worth remembering just how tiny the trade in lithium remained in a global context—it added up to just over 20,000 metric tons of lithium globally.[44] For every ton of lithium produced in 2005, the world used 100,000 tons of coal, 26,000 tons of steel, 13,000 tons of corn, and 65 tons of lead.[45] Unlike those resources, lithium was not treated as a global commodity. It was classified as a specialty chemical, sold on a contract basis, which required companies like SQM to tailor their product to the demands of buyers, matching the specifications for moisture levels, granularity, and the allowable levels of impurities that battery makers demanded. This process is known as certification. And though invisible to the end consumer, you can be sure each battery manufacturer knew where the lithium carbonate they used came from. As best I can tell, the lithium carbonate in my laptop likely originated on the salars of South America and, long before that, in the stars above.

COBALT: THE "BLOOD DIAMOND" OF THE LITHIUM-ION BATTERY INDUSTRY

Cobalt has been labeled the "blood diamond" of the lithium-ion battery industry.[46] It is an apt comparison. In the 1990s, a global campaign revealed that a sizable portion of the world's diamond trade originated in war-torn states in Central Africa, where diamonds were often extracted with forced labor (including that of children) in dangerous working conditions with the proceeds going to finance armed conflict. By the early 2000s, a concerted effort to certify the provenance of diamonds had alleviated much of the concern about blood diamonds.

But in 2010, the *New York Times* columnist Nicholas Kristof warned that "some of our elegant symbols of modernity—smartphones, laptops, and digital cameras—are built from minerals that seem to be fueling mass slaughter and rape in Congo."[47] Although Kristof focused attention on tantalum, tungsten, and tin, the so-called conflict minerals, he could have added cobalt to that list. It was a key component of the modern lithium-ion batteries powering almost every one of those devices. And, in the early 2000s, nearly 40 percent of the world's cobalt was sourced from the Katanga region of the Democratic Republic of the Congo.[48]

Compared to lightweight lithium, cobalt hails from a working-class district of the periodic table—the transition metals—where it is sandwiched between iron and nickel. Cobalt was initially valued as a pigment. When heated to 2,000°F, cobalt yields a vivid blue used to color pottery and glass. But for electrochemists, cobalt is most important when it is combined with oxygen to form a transition metal oxide. The ability to form such oxides is not unique to cobalt. Most transition metals can form oxides, which are used as pigments (titanium dioxide), chemical catalysts (to produce sulfur dioxide, for instance), and for electrochemical applications.[49]

In theory, a variety of transition metal oxides could serve in battery cathodes. Remember, this was John Goodenough laboratory's Nobel-worthy contribution to material science. The three potential cathodes his laboratory identified each combined lithium with a transition metal: cobalt, manganese, or iron. Of these, the lithium–cobalt oxide cathode was most common in consumer lithium-ion battery applications through the early 2010s. More recently, battery manufacturers have reduced the use of cobalt per battery, but with the scale of battery production continuing to grow, demand for cobalt has too.

Like every other material important to lithium-ion batteries, cobalt's story is not simple. Although cobalt is slightly more common than lithium in the Earth's crust, it is spread even more thinly. As a result, cobalt is usually recovered as a by-product of other minerals. After World War II, nickel refineries became a common source of cobalt, as demand for higher-quality stainless steel (which was an alloy of iron, chromium, and nickel) accelerated. To meet the demand for high-grade nickel, major nickel producers in Russia, Canada, Australia, China, and South Africa invested in hydrometallurgical refining processes that could separate nickel and cobalt as pure metals.[50] In the 1990s, these nickel operations met almost half of the world's cobalt production.

But most of the growth in demand for cobalt since the 2000s has been met by mines in the Democratic Republic of the Congo (DRC), which has increased production with the support of foreign, increasingly Chinese, investment in mining and refining capacity. The DRC exemplifies the "resource curse." The country is rich in minerals but poor and unstable. In 2000, the average lifespan in the DRC was forty-nine years, fewer than 10 percent of the population had access to electricity, and access to running water was rare. And the DRC was notorious for its weak governance. As one outside assessment put it in 2010, "petty and grand forms of corruption, as well as a complex web of political patronage permeate all sectors of the economy."[51]

The DRC's internal instability has long been driven by its rulers, both foreign and domestic, who plundered its wealth. This was the case in the late nineteenth century, when King Leopold II of Belgium ruthlessly stripped the country first of ivory and then rubber through a violent slave labor regime.[52] After the DRC attained its independence from Belgium in 1960, the Soviet Union and the United States jockeyed for influence, seeking to shore up the populous and mineral-rich ally at the height of the Cold War. The prize was the southern Katanga region, known for its stores of copper, cobalt, and strategically important uranium. The United States undertook covert operations to undermine the DRC's first elected leader in the early 1960s, a suspected Communist, and instead supported Mobuto Sese Seko, who forcibly seized power in 1965. Sese Seko ruled the country as a military dictator, enriching himself through an extensive patronage system that rested, in large part, on the country's mineral wealth.

Unrest in the DRC persisted into the twenty-first century, at the same time that the growing market for lithium-ion batteries began to drive a surge in cobalt demand. After Sese Seko's ouster in 1997, civil war wracked the country until a formal peace accord in 2002. That ushered in efforts to stabilize the DRC, keep the peace, and improve the country's welfare, largely by developing its mineral resources with the support of international investors, who had steered clear of the country since the 1970s. In 2002, to advance this strategy, the DRC adopted a new mining code developed by the World Bank, which aimed to draw international investment through a process for granting mining titles, favorable tax provisions, and the promise of regulatory stability for ten years. In this way, the DRC fit into the same pattern of neoliberalism important in Chile, in which resource-rich countries partnered with foreign investors, privatized investment, and developed export-oriented economies.[53]

During this time, cobalt mining expanded along two overlapping paths in the DRC. One was small-scale artisanal mining, undertaken with rudimentary processes both above and below ground. For many people in the DRC, artisanal mining offered an opportunity to earn money, especially as the country's economy struggled in the post–Sese Seko period. The government encouraged artisanal mining at some sites, recognizing its local economic benefits. But some artisanal mining also proceeded illegally. Those who participated in the work ranged from out-of-work mining officials to university teachers to unpaid government officials to migrants from neighboring countries to children. Estimates put the number of artisanal miners at between sixty thousand and one hundred thousand people in the early 2000s.[54]

Men generally descended into small, primitive mines in search of heterogenite, a dull brownish mineral rich in cobalt. The mines they worked were often dug solely with manual labor with no structural support, ventilation, or other safety measures. Landslides, suffocation, and drowning posed the highest risk of death. Women and children generally remained on the surface, sorting and washing the ore. Almost all worked without access to potable water or proper sanitation in mining camps where prostitution, disease, and ill-health were common. Few wore protective gear, such as face masks or gloves, to shield them from the long-term health risks associated with dust inhalation and heavy-metal exposure.

Although at least two artisanal miners' unions organized to represent miners' interests in the late 1990s and early 2000s, artisanal mining reflected the DRC's long history of corruption and patronage. In some cases, artisanal miners sought access to mining sites that had been contracted by the state to foreign developers. In other cases, local power brokers allied with the government regulated access to mining zones, levying taxes and fees in exchange for permission to mine, while controlling access to markets to sell the ore. In one case, the discovery of a seam of heterogenite in the small town of Kasulo in 2014 led miners to dig hundreds of small mines right in town, with some mine shafts starting inside peoples' homes.[55] Despite the hazards and corruption, outside observers concluded that the labor-intensive artisanal mining practices played a "crucial role" in stabilizing the Katanga region.[56]

But the role of children in the DRC's artisanal mining sector has been a long-standing and urgent concern. In 2009, investigators estimated that children under seventeen years of age accounted for nearly 50 percent of the workforce.[57] Although there is evidence some children work in the mines, most often they worked at the surface, sorting, cleaning, and transporting ore. Most children work because they need money to eat, to support their families, and to save money for school (local schools charge families a monthly fee to cover costs). In exchange for meager pay, children bore a heartbreaking toll. They endured grueling labor, chronic exposure to mining-related pollution, and, in the worst cases, maiming in mining-related accidents. Observers fault leading corporations for disavowing cobalt sourced with child labor while failing to undertake the due diligence necessary to certify their supply chains. In 2019, Washington, DC–based International Rights Advocates filed suit against Alphabet, Apple, Dell, Microsoft, and Tesla on behalf of thirteen children maimed, killed, and otherwise harmed while employed by cobalt operations that supplied the consumer electronics industry.[58]

The second path followed the DRC's efforts to draw international investment. Since the mid-1990s, when Sese Seko's regime began to falter, international mining companies and investors began circling, looking for opportunities to ally with rebels and gain access to the region. One of the biggest prizes was the Tenke-Fungurume copper-cobalt lode, claimed to be the largest undeveloped high-grade copper lode in the world. In 1996, a

FIGURE 3.2 In 2009 children made up approximately half of the workers engaged in artisanal cobalt mining in the Democratic Republic of the Congo. Courtesy of International Rights Advocates, Washington, DC.

Canadian-based firm, Tenke Mining Corporation, cultivated the country's leaders to win rights to a joint venture with the DRC's state-owned mining company, Gécamines, to develop Tenke-Fungurume.[59] When rebels seized power, Tenke Mining's $50 million down payment helped finance the rebels' final push to overthrow Sese Seko in 1997.[60] Between then and 2008, the project proceeded fitfully as civil war destabilized the region, and Tenke Mining sought an outside partner with deeper pockets to finance the mine.[61] Phelps Dodge, a US-based copper giant, filled that role starting in 2005 with a commitment to the $650 million mining project.[62]

Mining operations at the Tenke-Fungurume mine followed the template for modern, large-scale extractive industries. The concession measured 555 square miles in size, which included the mine site and enough land to create a "zero discharge" operation that self-contained all tailings, wastewater, and other by-products of mining and processing. The mine itself was a highly mechanized open-pit mining operation recovering ore that measured 2.1 percent copper and 0.3 percent cobalt by the end of 2009. On-site smelting processed the ore into high-grade copper and cobalt. In compliance with international lending requirements and the Congolese mining code, the company committed to hire local workers, improve infrastructure in neighboring villages, and mitigate environmental impacts.[63] The mine worked with local leaders to hire unskilled laborers from a list of local workers. According to the scholar Benjamin Rubbers, Tenke Fungurume employed 2,900 people in 2012, 62 of whom were expatriates.[64] The population in neighboring towns tripled as workers migrated to the region.

Making way for the mine meant displacing several villages and hundreds of people, closing off access to vast stretches of agricultural and grazing lands, and forcibly evicting hundreds of artisanal miners actively mining in the region (four of whom were reportedly killed in conflicts over the evictions in 2005).[65] Even after the industrial mine entered into production, hundreds of artisanal miners—many of whom had not gotten work at the mine—continued to press for access, mining covertly at night, staking their claim to what they considered their rightful lands.[66] In January 2008, five thousand protesters marched in Fungurume out of frustration with the mine development and lack of local support.[67] Humanitarian observers faulted Tenke Mining Company for its failures to follow through on its commitments to local investments and improvements in infrastructure. Although some projects were undertaken, "in none of these cases did the development of these projects take into account local realities, the real needs of the communities, or the future sustainability of these activities."[68] That did little to slow the Tenke Mining Company's development. By 2018, active mining pits stretched twelve miles from Tenke to Fungurume and accounted for 12 percent of global cobalt production.

The provenance of cobalt has been a growing concern internationally since the early 2000s. But that was not cobalt's only downside. It was also expensive. In 2000, the lithium–cobalt oxide cathode alone accounted for half the material costs of a battery cell.[69] When the price of cobalt nearly tripled in 2004, that forced some battery manufacturers to raise prices.[70] Cobalt is also toxic. Although cobalt forms an important component of the essential vitamin B_{12}, elevated exposure levels to cobalt pose health risks. Such exposures are most likely for workers processing cobalt or those living near cobalt mines. Children living in proximity to the cobalt mines in the Congo had cobalt levels in their urine that were more than forty times greater than average levels in the United States.[71] Despite these concerns, many portable consumer electronics sold since the 1990s have had ties to Congolese cobalt—cobalt contributed materially to the high performance of lithium-ion batteries.

Yet, cobalt is not essential to lithium-ion batteries: in the 1980s, Goodenough's lab identified three different potential cathodes—only one of which used cobalt in the cathode.

GRAPHITE: THE PHOENIX FROM THE ASHES

The performance of a lithium-ion battery depends on both the purity and the structural properties of the graphite anode. This was Yoshino's Nobel Prize–winning contribution to lithium-ion research at Asahi Kasei in the late 1980s. One of the key factors driving down the cost of lithium-ion batteries in the early 2000s was a shift away from the use of synthetic graphite toward lower-cost natural graphite, which was roughly one-fifth the price. But that shift also gave China a growing role in the supply chains important to lithium-ion batteries; it mined and produced almost all battery-grade natural graphite.

To understand China's outsized role in the little-known world of graphite, it is helpful to recall the alarm over China's efforts to restrict exports of another class of materials: rare earth elements. Rare earths are a family of seventeen elements valued for their exceptional magnetic and conductive properties. Although rare earths were generally not used in lithium-ion batteries, they did improve the performance and efficiency of related technologies, including electric motors, wind turbines, and lightbulbs. Rare earths also played a key role in military applications, including night-vision goggles and precision-guided missiles. China produced more than 97 percent of the world's rare earth elements. The *New York Times* broke the story in 2009, warning that "China is set to tighten its hammerlock on the market for some of the world's most obscure but valuable minerals" by reducing its export quotas.[72] When prices spiked, companies dependent on rare earths panicked, and the United States, European Union, and Japan filed a trade suit against China with the World Trade Organization.

China's dominance of the rare earths was not a product of geological serendipity. Rare earth elements, despite their name, are widespread across the globe. But mining and processing rare earths is an especially dirty and hazardous process, largely because many rare earth deposits, including those at China's largest mine in Bayan Obo in Inner Mongolia, co-occur with radioactive elements. Between the 1970s and early 2000s, the United States curtailed rare earth element production as environmental concerns and production costs mounted. Starting in the 1990s, as China turned toward the global market, it saw rare earths as an opportunity to promote development in its northern hinterland, strengthen its national economy, and gain a foothold in the global market. For two decades, China allowed rare earths to be produced with minimal protections for either human or environmental health—a strategy that drove down global prices, forced the closure of mines elsewhere, and gave China a near global monopoly on rare earths production.[73]

To many Western observers, it seemed that China was determined to leverage its control over rare earths for its own economic and geopolitical gains in 2010. But as the geographer Julie Michelle Klinger explains in her book *Rare Earth Frontiers*, that analysis dismisses China's domestic motivations for developing rare earths and the growing alarm in China over the industry's local consequences. Starting in 2004, residents of so-called cancer villages near the Bayan Obo mine raised concerns about air and water pollution that affected them most directly but also threatened the millions of downstream residents who depended upon the Yellow River. Contrary to Western allegations that the Chinese rare earth element industry was shrouded in secrecy, the Chinese state invested significant resources to better manage rare earth production in the 1990s and early 2000s, including documenting pollution, upgrading production facilities, and promoting research on rare earth applications. Thus, when China moved to reduce exports of rare earths in 2006, it was part of a broader shift in China's industrial minerals policy.[74]

It is this point that makes rare earth elements relevant to the history of graphite and lithium-ion batteries. China's changing role in the graphite industry mirrored its changing approach to the rare earths industry in the early twenty-first century, as it aimed to clean up the mining industry and encourage domestic value-added processing. Of the three major types of natural graphite—amorphous, flake, and vein—China dominated the production of flake graphite, which was the most important precursor for battery electrodes. To prepare flake graphite for use as an anode required purifying it to more than 99.9 percent carbon and mechanically processing the microscopic flakes into potato-shaped particles that more efficiently intercalate lithium ions. This process, developed in the early 2000s, results in what is called "spherical graphite."[75] Just how round and uniform these potato-shaped particles are plays a key role in determining how well a battery performs—how quickly can it respond to a power outage or accelerate an electric car?

Much as with the rare earths, China's dominance of the market for spherical graphite had more to do with its development-oriented industrial policy than it did with its geological good fortune. Flake graphite reserves are well distributed globally. But China scaled up its graphite industry starting in the 1970s by putting production ahead of public health and the local environment. In 2016, the *Washington Post* published an exposé highlighting the industry's impact on mining communities in northeastern China.[76] It was one of very few stories in the United States that covered the high environmental cost of graphite production. That concern would not have been news in China's Shandong province, however. Since the early 2000s, its residents had been urging the state to clean up a local graphite industry that had grown quickly to help meet the world's demand for graphite.

The most obvious form of graphite pollution is also surprisingly spectacular. Processing flake graphite into spherical graphite wastes 70 percent of the graphite. Some of that graphite escapes the mills and blows into the air. There it sparkles like glitter as the microscopic particles spin overhead. Once the graphite settles on the ground, however, it forms a layer of slippery silver-gray dust that coats everything, working through the windows and doors of homes and into the husks of corn and wheat.[77] Inhaling the small particulates poses a human health threat. More problematic, and less obvious, was the threat of water pollution, as one anonymous Chinese citizen warned in 2011.[78] China kept graphite production costs down by allowing water-based purification processes, instead of more energy-intensive and costly high-temperature heat treatments. Hydrometallurgical processing relied on a stew of strong bases and acids that resulted in surface- and ground-water pollution.

One of China's larger graphite ore bodies ran in an east-west arc through Shandong province, southeast of Beijing. High-grade graphite most often occurred in seams in metamorphic rocks such as quartzites and was located at the edges of tectonic blocks. Starting in the early 1990s, graphite mining in the region ramped up, as mining companies extracted ore from shallow open pits, using excavating machinery, and transported it onsite for initial processing and beneficiation. Although Shandong province accounted for only 5 percent of China's estimated graphite resources, it was one of three areas known for the high quality and large size of its graphite flakes.[79] Between the early 1990s and 2020, more than a dozen mining operations proliferated in a small area of Shandong province, rending the symmetrical fields of wheat and corn west of Pingdu and on the outskirts of villages such as Xishilingcun and Baoluocun.

Layering Shandong's graphite industry atop local farms became a persistent source of tension. Citizen activism prompted a crackdown on the graphite industry in the early 2010s. The regional newspaper, published online, reported that 40 percent of environmental complaints to local government officials targeted the graphite industry in 2013. Residents raised the alarm about mines and processing facilities that had expanded well beyond their permits, operated illegally at night to avoid oversight, drained ponds of wastewater into roadside ditches, and showered neighborhoods with graphite dust, soiling farm fields, homes, and their bodies.[80] Farmers explained that they could not sell their contaminated wheat or corn; not even their pigs would eat the soiled crops. One anonymous citizen urged the municipal government to "pay attention," especially to the water pollution stemming from improper disposal of hydrofluoric acid. When *Washington Post* reporters investigated the graphite industry in Shandong province, they and their contacts were followed and intimidated.[81]

The Pingdu Municipal Government, with the support of regional authorities, positioned itself as a forceful regulator of the graphite industry. It instituted a blanket shutdown of all graphite operations in December 2013. As the government explained, many graphite factories were older, smaller-scale operations that employed outdated processing technology with few pollution controls. Some operated illegally. The Pingdu Environmental Protection Bureau took the lead in cleaning up the industry. It undertook surprise inspections, required upgrades to pollution controls, and ran a public relations campaign to draw public attention and support.[82] It forced firms to remain offline until they met tighter pollution standards. The result was to shutter 20 percent of China's graphite production on environmental grounds at a time when demand was growing, raising concerns that prices could jump. Thus, what looked like a "supply squeeze" to industry observers in the West was linked to a sustained effort in China to clean up the production of graphite locally.[83]

The longer-term goal of the Chinese crackdown on the graphite industry was to consolidate graphite processing at larger-scale facilities located in newly designated industrial parks that would be engineered with tighter pollution controls and produce higher-value products. Such initiatives aligned with the national government's 2016–20 five-year economic development plan, which included closing "mining activities that use outdated techniques or are environmentally undesirable" and encouraging "green mining and the green mining industry."[84] In 2016, China established the Nanhai New District Graphene Industrial Park devoted to converting graphite into "ultra-high purity graphite and graphene materials."[85] This was the high-technology phoenix the Chinese envisioned rising from the ashes of the old graphite industry.

ASSEMBLING A LITHIUM-ION BATTERY

Up until the early 2000s, almost all lithium-ion battery manufacturing was concentrated in Japan, where companies such as Sony leveraged their strengths in factory automation to develop highly automated lithium-ion battery factories.[86] In some respects, this was surprising. American-based researchers had played key roles in pioneering lithium-ion battery technology since the 1970s. The United States had been the world's leading lithium supplier through the mid-1990s, and American companies still had significant investments in lithium processing both domestic and foreign. And, lastly, the United States had led the world in battery manufacturing, especially of lead-acid starter batteries and disposable batteries, for much of the twentieth century.

Ralph Brodd, a longtime battery industry executive and consultant, published a report in 2005 subtitled, "Why Are There No Volume Lithium-Ion Battery Manufacturers in the United States?" His answer boiled down to three points: First, major consumer battery manufacturers in the United States were locked into a competitive market selling AA, AAAs, and other disposable batteries in the 1990s. That made it hard to make the long-term investment needed to develop a lithium-ion battery industry. Second, the retail sales model that drove Duracell's and Energizer's sales (remember the Energizer bunny?) was fundamentally different from the market for lithium-ion batteries, which focused on direct sales to Asian consumer electronics manufacturers.[87] And, third, countries such as Japan, South Korea, and China provided strong state support for the advanced battery industry in the late 1990s and early 2000s, investing in basic research, providing subsidized loans, and encouraging development—well before the United States adopted similar policies.

The most consequential shift in the geography of the lithium-ion industry in the first decades of the twenty-first century was the industry's growing consolidation in China. In the early 2000s, much of the public attention in the United States focused on China's urban air pollution and its burgeoning coal industry. But that perspective overlooked China's early efforts to lay the foundations for a clean energy economy. Starting with China's Tenth Five-Year Plan in 2001, the state made advanced battery technologies, especially lithium-ion batteries, a priority under the "863 Program," which provided state-support for research and development. To accelerate manufacturing, the Chinese government offered subsidy and incentive packages that in some cases covered most operating and capital expenses for advanced battery manufacturers.[88]

By 2005, China's biggest lithium-ion battery producer was the home-grown BYD, Ltd. The company's founder, Wang Chuanfu, had trained and worked as a metallurgical chemist at the state-run General Research Institute for Non-Ferrous Metals in Beijing. At the age of twenty-nine, Wang saw an opportunity in China's underdeveloped battery industry. Wang left his government post to start BYD in 1995. He started the company in Shenzhen, China's first special economic zone slated for export-oriented development, located in southern China adjacent to Hong Kong. By the mid-1990s, Shenzhen was an exploding center of light manufacturing, drawing workers by the hundreds of thousands from rural China and producing a wide range of goods, including consumer electronics such as laptops, telephones, power tools, and toys, most of which needed a battery.[89] Wang pushed BYD into the global supply chain by positioning BYD as a supplier of low-cost rechargeable batteries to Shenzhen's consumer electronics industry.[90]

Wang expanded the company's production into lithium-ion batteries in the late 1990s. To do so, he followed the same strategy that had driven the company's growth since its founding: in place of expensive, largely automated production lines common at Japanese battery plants, BYD substituted carefully designed systems that leveraged human labor. A sixty-meter-long battery line at BYD, for instance, depended on the work of forty or fifty trained workers, seated along the assembly line, each repeatedly completing a very specific task.[91] Instead of operating production lines in an expensive humidity-controlled dry room, BYD limited dry rooms to key steps in the production process. Altogether, this strategy reduced BYD's capital costs, allowed it more flexibility (since it did not have large investments in expensive machines), and allowed it to take advantage of lower-cost labor in Shenzhen. BYD estimated its manufacturing strategy delivered a finished battery at 60 to 70 percent the cost of established manufacturers.[92] By 2002, BYD employed seventeen thousand people and produced two million batteries per day.[93] Wang became known as the "battery king."[94]

Although BYD has never had as high a profile as Foxconn, Apple's primary Taiwan-based manufacturing partner for the iPhone, BYD did draw scrutiny for its labor practices. In 2011, China Labor Watch described BYD as a "sweatshop" that relied on "never-ending waves of new workers that support BYD's cost-oriented manufacturing strategy." Working for BYD meant long hours, low pay, and living on a company-owned campus under strict rules. Based on China Labor Watch's research, nearly two-thirds of workers, most of whom were young women, had been at the company for less than a year—a figure explained in part by BYD's rapid growth but also its high turnover rate.[95] What China Labor Watch did not raise concerns about were occupational health risks—despite the intense working conditions, making lithium-ion batteries posed few health risks compared to manufacturing nickel-cadmium or lead-acid batteries.

BYD's rapid growth was emblematic of China's rise as a hub of export-oriented manufacturing for the West. But this was not just a story about China's growing manufacturing might or low-cost labor. During the early twenty-first century, with strong state support, China also positioned itself as the global leader in sourcing and processing the raw materials needed to support its manufacturing base, including those needed for lithium-ion batteries. In part, this was about China developing its domestic resources, such as the battery-grade graphite from Shandong province. China also produced some lithium from hardrock spodumene at several small mines concentrated in Sichuan province and from brine-based operations in far western China, making it the fifth-largest lithium producer in 2006. But to secure its position as a growing leader in lithium-ion battery manufacturing, the Chinese government eyed opportunities abroad.

China moved first to shore up its access to cobalt, securing a dominant interest in foreign investment in the Democratic Republic of the Congo. Chinese interest in the Congo was not new. Mobuto Sese Seko's regime had cultivated Chinese support and investment in the 1970s and 1980s, leading to a significant Chinese presence in the region. By the early 2000s, an estimated seventy Chinese-owned processing firms in Katanga province handled cobalt ore from Congolese miners, reselling to processing plants that were usually overseas and often Chinese-owned. These traders navigated tensions between locals, who resented the outsiders as "rough capitalists," and the Chinese, who often felt unfairly treated by allegedly corrupt Congolese officials.[96] Much of this trade was carried out by private Chinese firms with little support from the Chinese state. Starting in 2006, however, as China scaled its lithium-ion battery industry, it allied itself with the DRC's government to secure a foothold in the cobalt industry.

The most significant Chinese-DRC joint venture was the Sino Congolaise des Mines (Sicomines) project, negotiated between 2006 and 2009. In exchange for a 68 percent controlling interest in a copper-cobalt mine fifty miles east of the existing Tenke Fungurume mine, the state-sponsored Chinese consortium committed to investing approximately $6 billion in local infrastructure projects, including railways, roads, hospitals, solar installations, and housing. The Chinese financed the infrastructure projects, which were then to be paid off by the DRC's share of the mine's proceeds.[97] As at Tenke Fungurume, however, observers criticized the pace and quality of Chinese-financed infrastructure projects, which lagged behind implementation goals. When production started in 2016 at one-half capacity, however, Sicomines was already the third-largest cobalt mine in the DRC. That same year, another state-backed Chinese firm, China Molybdenum, acquired a controlling stake in Tenke Mining—which had theretofore been largely owned by Canadian and US investors.[98]

The story was much the same in the case of the lithium industry. By 2011, China already had the capacity to produce approximately fifty thousand metric tons of lithium carbonate per year.[99] As China began to position itself as a leader in lithium-ion production globally, it moved aggressively to secure access to lithium resources. In the early 2010s, China's leading lithium processors began to make major investments in overseas lithium production and enter into partnerships to locate more lithium processing in China. In 2010, Ganfeng Lithium invested in International Lithium, which was developing a brine-based operation in Argentina. In 2011, China partnered with Australia-based Galaxy Resources to expedite construction of a lithium carbonate plant in Jiangsu province in China. And, in 2012, China's Chengdu Tianqi Industry Group acquired a controlling stake in Australia-based Talison Lithium, which would soon become the world's largest producer of lithium. In short, China was moving to vertically integrate the lithium-ion battery industry, from mining to chemical processing to battery manufacturing, and according to industry observers, they had a significant cost advantage at each step of the process.[100]

The growth of companies like BYD and Chinese manufacturing processing capacity along the entire lithium-ion battery supply chain helped drive the growth of the lithium-ion battery industry in the early twenty-first century. Between 2000 and 2010, the annual production of lithium-ion batteries grew twelvefold, accounting for almost all the growth in the consumer rechargeable battery industry. During the same time period, the price of the 18650 battery cell fell by a factor of ten, largely driven by increasing economies of scale in sourcing materials and in manufacturing.[101] Increasingly, the world's lithium-ion batteries were being made in China. Even when they were being manufactured in Japan or South Korea, much of the lithium, cobalt, separators, and other materials were processed in China first. To some observers, China's advantage in consumer electronics signaled more than just a global shift in manufacturing. It gave China a head start in the race to a clean energy future.

THE EXPLODING BLACK BOX

To the extent consumers paid much attention to lithium-ion batteries in the early 2000s, however, it was likely because of reports about battery fires. There had been sporadic issues with overheating cell phones and laptops in the 1990s, but as the use of lithium-ion batteries accelerated in the early 2000s, the problems mounted. The problem drew public attention in 2006, when laptop computers began catching fire in offices, airports, hotel rooms, and people's homes. One of the most dramatic incidents happened in rural Nevada in July 2006. A Dell laptop left in a pickup truck caught fire, igniting ammunition stored in the glove box, which then caused the gas tank to explode, sending the entire truck up in flames. The two men loading the pickup truck were unharmed.[102] In August 2006, Dell recalled 4.1 million laptop batteries on account of the fire risk. Apple followed suit, recalling 1.8 million laptop batteries. All of the batteries were manufactured by Sony. And in September 2006, in cooperation with the US Consumer Product Safety Commission, Sony initiated a worldwide recall of the lithium-ion battery packs it had manufactured for use in notebook computers since 2005.

It was surprising that the faulty batteries were manufactured by Sony, for much of Sony's pioneering work leading up to the introduction of the lithium-ion battery in 1991 had centered on ensuring their safety. Unlike older battery chemistries, lithium-ion batteries employed a flammable electrolyte, which could ignite in the event of overheating, a short circuit, or structural damage. With twice the voltage of most other chemistries, a short circuit would generate more heat too. Sony had already adopted several safety features important to lithium-ion batteries. If pressure built within the battery, a flexible aluminum "burst disk" was designed to bulge outward, thereby breaking the charging circuit. If the battery overheated, the microscopic pores in the plastic separator between the anode and cathode melted shut, stopping the chemical reaction. And other temperature-sensitive conductors shut down if the battery reached its "trip temperature." But risk still remained: if a battery became too hot, the separator could fail entirely, putting the anode and cathode in direct contact. Engineers warned that would result in "vigorous heat generation."[103]

As the pressure to produce lower-cost, higher-capacity batteries mounted in the 2000s, one of the easiest ways to gain an advantage was to make separators even thinner. That strategy reduced the margin for error in manufacturing. The problem for Sony in the early 2000s was contamination at one of its lithium-ion battery plants. Based on its own internal analyses, Sony determined that metallic particles produced during the manufacturing process were contaminating battery cells and, in rare circumstances, creating the potential for a short circuit. But, Sony explained, the probability of an accident "very much" depended on how the battery pack was integrated into the laptop. As the historian Matthew Eisler has pointed out, an enormous gulf separated the engineers driving product development at companies like Apple and Dell and the battery chemists at companies such as Sony, Panasonic, LG Chem, or BYD in the 1990s and 2000s. It was both an intellectual gulf, in that electrical engineers knew few of the details of battery chemistries and proper power management, and it was a geographical gulf, as the trend toward outsourcing manufacturing in the electronics industry accelerated in the 1990s, making battery packs just one more component to be optimized for high performance and low cost.[104]

The recall of consumer laptops was the largest consumer product recall in history at the time—it included ten million batteries and cost Sony $500 million to address. As Eisler has explained, the incident was a reminder that despite intensive safety engineering and careful manufacturing, sometimes the "lithium genie gets out of the bottle, with potentially catastrophic consequences."[105] Indeed, there has been a steady stream of issues and recalls since, including for additional laptops, smartphones (like Samsung's top-of-the-line Galaxy Note in 2016), toys (such as hoverboards in 2017), and electric cars (such as the Chevrolet Bolt since 2019). Vaping devices have been especially problematic. They accounted for two thousand emergency room visits between 2015 and 2017 and two deaths since 2018 in the United States alone.[106] And lithium-ion batteries have posed risks during manufacturing (several lithium-ion battery factories have been shut down temporarily due to fires) and transport (at least three airline incidents are attributed to shipments of lithium-ion batteries, including a fatal crash of a UPS cargo plane in 2010).

Such incidents are sobering. Yet, considering that billions of lithium-ion cells were in use, such incidents remained relatively rare. For most people, the only frustration with lithium-ion batteries was that, despite their advantages, they still did not charge fast enough or last long enough. But the safety concerns that surrounded lithium-ion batteries in 2006 had one important consequence for the industry. With laptop batteries going up in flames, governments issuing recalls, and lithium-ion batteries making headlines, many automakers hesitated to bet that the same technology that had enabled the "wireless revolution" could power a transition to electric cars. In their view, lithium-ion batteries remained too expensive and too risky. Indeed, the early hybrid cars in the 1990s and 2000s, such as the Toyota Prius, all relied upon lower-performing, but safer, nickel–metal hydride batteries. There was one exception. In July 2006, a month before the US Consumer Product Commission forced Sony to recall millions of lithium-ion laptop battery packs, a little-known California-based start-up company, Tesla Motors, unveiled its first vehicle: an electric sports car it dubbed the Roadster, which harnessed Japanese-made 18650 lithium-ion battery cells—the kind most often used in laptops or power tools—into a carefully engineered, liquid-cooled battery pack capable of powering the sports car for more than 200 miles.

What stood out most about the Nobel Prize Committee's awards ceremony for the researchers who pioneered the lithium-ion battery was how much the affair looked to the future instead of the past. Although the award recognized Whittingham's, Goodenough's, and Yoshino's pathbreaking research contributions, an overarching theme of the December 2019 ceremony was the prospect of a bright, lithium-powered future. In its commendation, the Nobel Prize Committee emphasized that lithium-ion batteries "have laid the foundation of a wireless, fossil fuel-free society, and are of the greatest benefit to humankind." The laureates themselves echoed that anticipatory optimism, describing how lithium-ion batteries had already enabled a wireless revolution and promised a renewable energy future that would "liberate" humans from fossil fuels and make it possible for humans to create "a sustainable society."[107]

Although such a world would be far preferable to a world powered by fossil fuels, what none of the laureates or their Swedish hosts emphasized was that making this transition would mean mobilizing a wide array of metals and chemicals, each of which had a cascade of consequences for environments and communities around the world. Thus, to begin to take measure of the implications of scaling up a clean energy future powered by lithium-ion batteries, it is necessary to consider how the world met the growing demand for the lithium, cobalt, and graphite important to powering the wireless revolution in the 1990s and early 2000s. As we have seen, enabling the wireless revolution at the start of the twenty-first century was a process deeply grounded in local landscapes and communities in places as far-flung as Chile, the Democratic Republic of the Congo, and China.

On the one hand, the costs of this wireless revolution were, relatively speaking, small compared to other systems of materials production. The total mass of lithium-ion batteries manufactured in 2010 added up to about five million tons. Even if we could fully factor in all the slag, waste materials, water, and energy that went into manufacturing those five million tons of batteries, the impacts would hardly amount to a rounding error compared to the consequences of the steel industry, which manufactured 1.4 billion tons of steel the same year. Although the scale may have been relatively small, the lithium-ion battery industry's reach was large and growing. And by the 2010s, as prospects for a clean energy future began to accelerate, so too did the scale of lithium-ion battery production. While a smartphone needed a battery that could store 10 watt-hours of energy, an electric car demanded a battery that could store at least a thousand times more energy. As one industry observer commented, "The battery megafactories are coming."[108] That set off a race to build the battery factories, find the raw materials, and strengthen the supply chains on which a clean-energy future will turn.

4

ELECTRIC CARS, TESLA, AND
A ZERO-EMISSIONS FUTURE

IN FEBRUARY 2014, TESLA MOTORS ANNOUNCED ITS PLAN TO BUILD A
massive "Gigafactory" to manufacture batteries for its electric cars. At the
time, global lithium-ion battery manufacturing capacity for electric vehicles
added up to 28 gigawatt-hours annually, with almost all of that capacity concentrated in Asia.[1] Tesla planned to build a factory in the American South-
west that would start production by 2017 and ramp up to 35 gigawatt-hours
of battery manufacturing capacity by 2020. One industry analyst commented, "It is the most exciting thing to happen in the auto industry since
Ford went public in the 1950s."[2]

Tesla and its chief executive, Elon Musk, emerged as the rising stars of
an electric car revolution in the early 2010s. The Gigafactory was a key step
in advancing Tesla's "master plan." The company's goal was to expedite the
transition from a "mine-and-burn hydrocarbon economy" to a "solar electric
economy." It aimed to leverage sales of a high-end sports car to finance development of a luxury sedan, which would, in turn, generate revenue to develop
an affordable, mass-market electric sedan. By 2013, its master plan was well
underway. The company had delivered the Roadster sports car in 2008 and
the Model S luxury sedan in 2012. In truth, each new model did not so much
finance the next model as build confidence in Tesla's strategy, which allowed

the company to continue raising the capital needed to finance its growth, including nearly $2 billion for the first Gigafactory.[3]

In 2013, Tesla sold roughly twenty-two thousand cars. That was a fraction of the ten million cars General Motors sold that year. But putting Tesla's electric cars on the road required nearly half as much lithium-ion battery capacity as did the one billion smartphones manufactured that year. Looking ahead, Tesla warned that both the short supply and high cost of batteries would be a limiting factor in its ability to bring a mass-market electric vehicle to the market. The Gigafactory aimed to solve both of those problems. Tesla projected it would drive down its battery costs by 30 percent, and, at full scale, it would manufacture enough batteries annually to power five hundred thousand long-range electric cars.[4]

In the early 2000s, the lithium-ion battery industry had grown exponentially, as it powered the wireless revolution. Beginning in the 2010s, the industry began to retool in anticipation of a new market for electric cars. Unlike with laptops, cell phones, and other consumer electronic products, for which the lithium-ion battery served as an enabling technology—making devices more portable, powerful, and long-lasting—electric cars were different. In the case of the automobile, the battery-powered electric car needed to displace an existing technology, the internal combustion engine, that most people found satisfactory on a day-to-day basis.

This chapter offers a brief history of how the lithium-ion battery industry positioned itself in the 2010s to support the emerging electric car revolution. For those who see in the challenge of climate change an urgent need and opportunity to remake the world order, challenging the underpinnings of capitalism, unraveling global supply chains, and shifting power from corporations back to local communities, this chapter will offer little of interest. Although companies like Tesla and competitors, such as China-based BYD, aimed to break with a twentieth-century fossil-fueled economy, they did not break with a global economic order predicated on growth, intensive resource extraction, large-scale manufacturing, and consumerism. This chapter is organized around three points, as it surveys the recent history of the battery-powered electric car and prospects for a clean-energy future.

First, despite Asia's and especially China's head start in lithium-ion battery manufacturing, the future of the industry was effectively a blank slate in 2010. Less than 5 percent of the advanced battery manufacturing capacity that would come online by 2020 and less than 1 percent of the capacity needed by 2030 had been built. That set off a geopolitical scramble, as the United States, Europe, and South American countries sought to gain a toehold in the industry. Industry analysts made clear that China's growing lead in lithium-ion battery production was not simply a product of low labor costs—labor actually accounted for a fraction of the cost of battery production by 2010. Instead, it was a product of a coordinated set of government policies that supported clean energy research and investment, encouraged resource development and raw materials supply chains, and incentivized private investment in manufacturing. Starting in 2009, on the heels of the Great Recession and with concerns about climate change mounting, the Obama administration made supporting manufacturing capacity for advanced batteries and electric cars a centerpiece of its efforts to position the United States at the forefront of a clean energy transition. Tesla became the unlikely headliner in the administration's efforts to vault the United States to the forefront of a clean energy economy.

Second, were electric cars really sustainable? In the 2010s, proponents of a clean energy future emphasized the key role electric cars would play in a "zero emissions" future. Transportation accounted for one-quarter of US greenhouse gas emissions. Although electric cars could help reduce emissions from burning gasoline, especially if they were charged using low-carbon sources of electricity, their role in addressing climate change was not so simple. As became clear in the early 2010s, before an electric car hit the road, it already had a larger carbon footprint compared to a conventional car, on account of the extra materials and energy needed to manufacture the battery. Tesla saw in the new scale and location of the Gigafactory an opportunity to improve the sustainability of its supply chains and, ultimately, its vehicles. Tesla depended on Panasonic for its battery cells, which were manufactured in Japan, using materials sourced from around the world. By localizing production in Nevada and sourcing more materials domestically, Tesla aimed to lower costs and curb the impacts of manufacturing lithium-ion batteries. Tesla and Panasonic entered into a joint partnership, with Panasonic manufacturing lithium-ion battery cells on-site at the Gigafactory. In doing so, Tesla aimed to leverage the Gigafactory's scale to support new sources of raw materials in North America.[5] Tesla also planned to power the Gigafactory entirely with renewable energy sources. Between sourcing materials locally and using renewable energy, the Gigafactory promised to substantially reduce the upfront environmental impact of manufacturing lithium-ion batteries. And Tesla saw the Gigafactory as the hub of a closed-loop industrial ecosystem, recycling the materials from scrapped and retired lithium-ion batteries to make new ones.[6]

Third, the Nevada Gigafactory announcement gave new life to an old concern among electric vehicle proponents: could supplies of lithium, cobalt, graphite, and the other materials important to an electric car future be scaled up rapidly enough to meet Tesla's and other companies' electric car ambitions? Analysts anticipated that the Gigafactory would become a disruptive force in raw materials markets— driving new concerns about resource limitations and spurring investments in resource extraction around the world. Indeed, for all the public attention that the Gigafactory attracted, by the mid-2010s, it was just one of many large-scale lithium-ion battery factories being planned globally. What few of those projections accounted for, however, were the changes in the chemistry of lithium-ion batteries. Starting in the 2010s, electric car companies began to deploy newer lithium-ion battery chemistries that used more nickel—which allowed for longer-range vehicles—and less cobalt, which remained both expensive and ethically problematic. That meant not only that the scale of materials demand grew but also that the composition of that demand changed, reshuffling the burdens of resource extraction for local communities around the world that supplied the lithium, cobalt, nickel, and other materials increasingly important to a clean energy future.

THE PRIUS EFFECT

In the mid-1990s, the Rocky Mountain Institute—the clean energy think tank cofounded by alternative-energy champions Hunter and Amory Lovins—envisioned what they described as the "hyper car"—a lightweight, fuel-efficient, mass-produced, and software-driven vehicle that could be five times more fuel-efficient than existing vehicles. They warned that while the auto industry was "painstakingly refining designs," advances in materials, software, motors, energy storage, and manufacturing had the potential to transform the auto industry, "perhaps with terrifying speed."[7]

In 1990, the California Air Resources Board adopted a far-reaching set of policies aimed at accelerating the transition to more efficient cars and alternative-fuel vehicles, including zero-emissions vehicles. The combination of heavy traffic, suburban sprawl, and an urban geography hemmed in by coastal mountain ranges and inland valleys meant that cities like Los

Angeles and Sacramento could be shrouded in acrid smog for weeks on end. In 1988, parts of California violated federal air pollution standards every other day on average.[8] California mandated that by 1998, 2 percent of cars sold in California be zero emissions, expanding to 10 percent by 2003. They hoped the policy would "spur a multi-billion-dollar race to bring alternative fuel and electric vehicles to market."[9]

In the early 1990s, the future of the car seemed wide open. Between California's regulatory ambitions and the newfound enthusiasm of the Clinton administration, the United States tried to put automobile innovation on the fast track. Starting in 1991, the United States Advanced Battery Consortium brought together the Big Three automakers—Chrysler, General Motors, and Ford—to "pool their technical knowledge and funding to accelerate" the development of batteries for electric vehicles in the hope of meeting California's zero-emissions mandate. Two years later, the Clinton administration launched the Partnership for a New Generation of Vehicles, which spurred the Big Three automakers to develop an advanced, no-compromise "super car" based on emerging gasoline-hybrid technology that got 80 miles to the gallon.[10]

But these initiatives did little in the near term to build an advanced battery industry in the United States or, in the case of the "super car," to move beyond a concept vehicle. Automakers put a few thousand battery-powered compliance cars on the market in the late 1990s, including a Toyota RAV4, a Ford Ranger pickup truck, and the short-lived General Motors EV1. But these vehicles all had ranges of fewer than 100 miles, relied on finicky lead-acid or later nickel–metal hydride batteries, and were expensive. Despite the enthusiasm of early drivers, the vehicles were generally available only in limited numbers to satisfy California regulations. GM became notorious for "killing" the EV1 in 2003, much to the frustration of electric car enthusiasts. Enthusiasts even staged a funeral for the car at a Hollywood cemetery—an event that figured prominently in the 2006 documentary *Who Killed the Electric Car?*[11]

While the automobile industry invested only half-heartedly in a new generation of cars, they seemed to put more of their efforts into challenging California's zero-emissions mandate as too costly and limiting. In their view, alternative technologies could more economically and effectively meet California's goals of reducing air pollution. In 2003, the year that California had hoped 10 percent of new car sales would be electric, the state instead rolled back the requirement, allowing manufacturers to count sales of high-efficiency cars and hybrids toward the goal. In the early 2000s, George W. Bush's administration killed the Clinton administration's "super car" in favor of a "FreedomCAR" program that prioritized long-term research into hydrogen-powered fuel cell vehicles—an initiative that posed little threat in the near term either to the auto or oil industries.

With gasoline prices at historic lows in the late 1990s and early 2000s and profits from sport utility vehicles and pickup trucks rolling in, electric cars and other high-efficiency vehicles remained sideshows at best for American automakers. Enter the Toyota Prius. With the tagline, "Eat my voltage," the Toyota Prius debuted in Japan in 1997 and hit the US market in July 2000. Toyota promised that the hybrid's combination of a high-efficiency gasoline engine combined with an electric motor and battery meant: "No recharging stations. No plugs. No compromises." The Prius relied on a battery to capture energy during braking, which it could then use to power the vehicle while idling and driving at low speeds and provide power assist to the conventional gasoline engine at high speeds. It got up to 52 miles per gallon, more than double that of the top-selling Toyota Camry.[12] Despite its high cost and modest performance, it was the first mass-produced car to offer an alternative to the conventional internal combustion engine since the 1920s.

The Toyota Prius became a surprise success. Celebrities plugged it at Earth Day celebrations in April 2000. The Sierra Club took it on a fuel-efficiency road tour around the United States. The second version topped the Consumer Reports survey for owner satisfaction in 2005.[13] It was the eighth-best-selling car in the United States in 2007.[14] And by 2011, Toyota had sold one million hybrids in the United States and three million worldwide.[15] With its distinctive styling, heightened by the introduction of a second-generation hatchback in 2003, the Prius became a badge of environmental consciousness and, to many, a symbol of the future of automobility. *Car and Driver* described it as the first car that ran on "guilt."[16] In fact, the Prius relied on a nickel–metal hydride battery that stored roughly 1.3 kilowatt-hours of electricity, an electric motor, and a small gasoline engine working alternatively or together to power the car. Toyota chose a nickel–metal hydride battery, instead of the more energy-dense lithium-ion batteries driving the consumer electronics industry, because it was safer, more durable, and less expensive.

For those concerned about the environment, the Prius offered a tangible step toward a greener and more fuel-efficient future. Indeed, in the first decade of the twenty-first century, a drumbeat of concern about global warming began to build, amplified by Hurricane Katrina in 2005, Al Gore's documentary *An Inconvenient Truth* in 2006, and the fourth major international report on climate change in 2007, which stated that "warming of the climate is unequivocal" and that human activity was "very likely" driving the change. Although it is surprising in retrospect, given the toxic politics of climate change in the 2010s, in the 2000s there was substantial bipartisan support for action on climate change. President George W. Bush's administration refused to take a lead on the issue in the early 2000s, throwing its lot in with the fossil fuel industry, but by 2008 support on both sides of the aisle for climate policy grew. Republican leader Newt Gingrich and Democratic leader Nancy Pelosi sat side by side for a political advertisement urging bipartisan climate action that year. And soon-to-be-president Barack Obama and his Republican challenger, John McCain, both made federal action on climate change a priority in their presidential campaigns.[17]

The Prius made clear just how far behind the United States had fallen in the race to invent the future of the car. While the Big Three American automakers raked in profits from pickup trucks and SUVs in the 1990s and early 2000s and fought the California regulations, Japanese firms outpaced the United States in registering patents related to hybrid and electric automobiles. Companies like Toyota and Honda were the only ones bringing hybrids to the market. To do so, they were using Japanese-manufactured batteries and electric motors. When Ford decided to launch its first hybrid cars in 2004, it licensed the technology from Toyota. When GM launched its first long-range hybrid vehicle in 2010, the Chevrolet Volt, it turned to LG Chem, a South Korean company, for the battery packs. All of this became a growing concern as the American auto industry's bet on big cars began to sour. By 2008, high gasoline prices, collapsing sales, a decade of layoffs, and a global financial crisis threw the American auto industry into an unprecedented crisis—the Big Three automakers were losing tens of billions of dollars annually, closing down factories and putting employees out of work, and teetering on the edge of bankruptcy.[18]

The newly elected Obama administration threw a lifeline to the American auto industry in 2009. Building on President Obama's campaign promises to revitalize the American manufacturing sector and economy by making a massive investment in clean energy, the administration leveraged its $80 billion bailout of GM and Chrysler to open up negotiations that led to higher fuel-efficiency standards for vehicles—increasing from 27 to 34 miles per gallon by 2016.[19] The administration targeted $2.4 billion from the American Recovery and Reinvestment Act of 2009, the stimulus act that aimed to right the economy after the 2008 fiscal crisis, to support investments in a supply chain for advanced batteries and vehicles in the United States. And it channeled additional stimulus funds through the Advanced Technology Vehicles Manufacturing Loan Program to domestic manufacturing of more energy-efficient vehicles.[20]

All together, these initiatives were designed to support President Obama's broader goal that the United States "lead the world in building the next generation of clean cars."[21] To spur the industry forward, the Obama administration set a lofty goal: putting one million electric vehicles on the road by 2015. To bolster consumer demand, the stimulus act provided buyers of plug-in vehicles up to $7,500 in federal tax credits. Obama did not want consumers just claiming tax credits for new hybrids made in Japan. His administration targeted these investments to jump-start advanced vehicle manufacturing in the United States. Countries like Japan and China already had a head start in manufacturing solar panels, wind turbines, and advanced batteries. For a century, auto manufacturing had been a pillar of the American economy. Without sustained investment, President Obama warned that the United States risked ceding the future of the auto industry to competitors too.[22] In the throes of a global economic crisis, the future of the electric car suddenly flickered to life, even in the United States.

JUMP-STARTING THE ELECTRIC CAR INDUSTRY

The burst of federal stimulus spending set toes tapping in anticipation all along the lithium-ion battery supply chain. While the Department of Energy's Advanced Technology Vehicles Manufacturing Program provided the loans to automakers, such as Nissan, GM, Fisker, and Tesla, the 2009 stimulus act provided investments that were meant to lay the groundwork not just for electric cars but also for the industrial ecosystem needed to support them. In the early 2010s, the Department of Energy channeled stimulus funds into domestic lithium-mining operations in Nevada, next-generation battery research and development at universities and national labs, subsidized loans for battery manufacturers, pilot programs to deploy electric vehicles and the necessary charging infrastructure, and research on lithium-ion battery recycling. In short, leveraging the stimulus funding, the Obama administration and the Department of Energy aimed to spur electric vehicles forward, from cradle to grave. In practice, stimulus funding largely went to battery assembly, vehicle manufacturing, and technology deployment (such as charging stations or subsidizing car purchases)—not securing raw materials such as lithium or graphite or developing the capacity to refine them into battery-quality materials. By 2011, the federal government had committed more than $2 billion in stimulus act funds to the battery industry and loaned $2.4 billion through the Advanced Technology Vehicles Manufacturing Program to scale up electric vehicle production.[23]

A123 Systems featured prominently in the Obama administration's clean energy initiative. A123 had emerged out of Yet Ming Chiang's material science and engineering laboratory at the Massachusetts Institute of Technology in 2001. By 2006, the private start-up had already demonstrated its viability. It manufactured its proprietary lithium-ion battery cells in South Korea and China to supply Black & Decker's line of DeWalt power tools. But the real goal was to power electric cars. To showcase the potential of his company's technology, Chiang loved to show off his souped-up Toyota Prius, which had been retrofitted with A123's advanced lithium-ion batteries to double its efficiency.[24] In 2009, A123's prospects soared. In September, A123 went public, raising $400 million. It secured a $250 million grant from the Department of Energy for a new battery plant in Michigan. It was in talks with GM and Fisker Motors about supplying batteries for their new electric cars. And, in 2010, President Obama celebrated the opening of A123's advanced battery manufacturing facility in Michigan. "Made in America" batteries, he crowed. "That's what you guys are helping to make happen."[25]

A123's prospects seemed to accelerate when it secured a contract to supply the batteries for Fisker Motors' first vehicle, the Karma. Although little remembered now, Fisker captured nearly as much attention as did Tesla after its founding in 2007. In 2008, Fisker unveiled a luxury plug-in hybrid vehicle with a projected electric range of 50 miles, which could be extended by 250 miles or more with an onboard gasoline generator.[26] Like A123, Fisker Motors was riding the wave of public-private investment in an electric car future. By 2009, it had secured $500 million in venture capital investment and $500 million in support from the federal government—despite having not yet manufactured a car. "We're making a bet on the future," promised Vice President Joe Biden in October 2009, speaking at a shuttered GM car factory in Delaware, which was slated as the future site of Fisker Motors. Fisker promised to reopen the plant as an electric car factory and manufacture 75,000 to 100,000 plug-in vehicles per year by 2014.[27]

For proponents of an electric car future, suddenly everything seemed to be happening right on schedule. In March 2010, Tesla launched the Model S, its first mass-production, long-range all-electric sedan. That December Nissan released its all-electric Leaf with a range of 100 miles, and Chevrolet released the Volt, which could travel 40 miles on battery alone and another 375 miles powered by an onboard gasoline generator. In 2013, Nissan began manufacturing the Leaf for the North American market at its Smyrna, Tennessee, factory (which had been retooled using funds loaned by the stimulus act). And a federally funded EV Project supported the installation of what would amount to more than twelve thousand charging stations in US metropolitan areas by 2013.[28] With concerns about global warming mounting and gasoline prices rising, the Obama administration prepared even tighter rules that would more than double fuel-efficiency standards for passenger vehicles. Electric cars seemed primed to merge into the mainstream of American car culture.

IS GOING ELECTRIC WORTH IT?

In 2013, Bjorn Lomborg, a European-based economist who had made a name for himself as the "skeptical environmentalist" (the title of his controversial 1998 book), published a column in the *Wall Street Journal* titled "Green Cars Have a Dirty Little Secret." In it, he pointed out something that any reader of this book already knows: electric cars are not really "zero emissions." But at a time when the Obama administration was putting electric cars at the center of its clean energy stimulus program and the likes of Leonardo DiCaprio touted their new electric cars as the solution to climate change, Lomborg was eager to point out that electric cars were not "truly green." It was a message in which the conservative editorial board of the *Wall Street Journal* likely delighted and that probably came as a surprise to many readers.[29]

Since the late 1990s, environmental groups, policymakers, and activists had been championing "zero-emissions" vehicles as key to a sustainable future. In 2012, the Cambridge, Massachusetts–based Union of Concerned Scientists published its first white paper assessing the environmental impacts of electric cars. Its message was clear: "electric vehicles—coupled with clean and sustainable electricity—are important parts of the solution" to climate change. That was clearly so in a state such as Vermont or Oregon, where hydroelectricity and nuclear power dominated the electrical grid. But their analysis considered whether it still made sense to drive an electric car in

Wyoming or West Virginia, where coal-fired power plants supplied the electrical grid. Their analysis showed that even in states with dirtier electrical grids, an electric car was still preferable to a conventional car, although not to a high-efficiency hybrid vehicle.[30]

But Lomborg found two faults with this reasoning. First, as an economist, he was especially concerned with cost. In his analysis, an electric car avoided about nine tons of carbon dioxide over a decade. Yet, one could easily buy carbon offsets or other credits that added up to nine tons of avoided carbon dioxide emissions—for instance, by planting trees. In his estimation, those alternatives would cost about forty-eight dollars. With the federal government providing a $7,500 subsidy for electric car purchases and billions of dollars in stimulus funding to support companies like Nissan, Tesla, and Fisker, Lomborg noted that made for a "very poor deal for taxpayers."[31] It was a seemingly reasonable point, but it made two key assumptions: first, that carbon offsets meaningfully reduced the atmospheric levels of carbon dioxide (a point of substantial debate) and, second, that the most immediate value in reducing greenhouse gases were the near-term costs and benefits. But proponents of an electric car future measured the value of those tax credits and subsidies not in terms of their value but in terms of their potential. As Obama had made clear, the federal support for electric cars was but one piece of a broader strategy aimed at building a clean energy economy. And in that future, reducing emissions from transportation was going to be key. In 2013, transportation still accounted for one-quarter of US greenhouse gas emissions and remained the fastest-growing source of emissions.[32]

But that analysis omitted the impacts of manufacturing electric cars in the first place. What if the extra emissions from manufacturing an electric car, with its large battery, actually overwhelmed the reduced emissions from driving electric? This was Lomborg's second point. Without quibbling over whether the life-cycle analyses that Lomborg relied upon were exactly right and ignoring the obvious glee he took in his role as the skeptic, environmentalists had a bad habit of focusing on how things were used and disposed of rather than on how things were sourced and manufactured in the first place. For instance, that the Toyota Prius's high-efficiency drivetrain was the largest end-use for rare earth metals in the 2000s drew little attention at the time. The Union of Concerned Scientists fell into a similar trap. Their first report on electric cars disregarded the upfront impacts of manufacturing electric cars entirely. They were hardly alone. The federal government, environmental groups, and many industry analysts focused almost all of their attention on charging cars, not making them. For instance, the federal fuel-economy stickers adopted in 2011 included a new category labeled "Environment" that ranked vehicles based solely on their "tailpipe emissions," purposefully focusing on the "vehicle itself, rather than the broader system."[33]

That was problematic for two reasons. First, in areas reliant on coal-fired power plants, the benefits of an electric vehicle were marginal, as the Union of Concerned Scientists had already shown in their report; an electric car reduced emissions between 11 to 33 percent compared to a conventional car. Second, life-cycle analyses estimated that once the additional materials and energy needed to manufacture an electric car battery were factored in, the upfront impacts of manufacturing an electric car were twice that of a conventional car. This was a point the Union of Concerned Scientists report omitted, despite a growing number of scholarly studies on the environmental impacts of manufacturing electric cars.[34] For instance, by the time an electric car like the Nissan Leaf hit the road, manufacturing it produced twice the greenhouse gas emissions of a comparable conventional car.[35] Of course, operating the Leaf resulted in far fewer greenhouse gas emissions than did driving a conventional car. But based on initial analyses, it still took years to pay off the upfront greenhouse gases incurred during sourcing and manufacturing.

In 2015, the Union of Concerned Scientists released an updated and expanded report on the "cradle to grave" impacts of clean cars that included an estimate of the impacts of manufacturing lithium-ion batteries.[36] Their analysis was based upon work at the Argonne National Laboratory, which had been refining a publicly available model for assessing the complete life cycle of vehicles since the late 1990s and a growing scholarly literature on the life cycle of electric vehicles. In the Union of Concerned Scientists' expanded analysis, even a full-sized battery electric vehicle charged in coal-reliant Wyoming paid off its upfront carbon debt within three years. In their assessment, while the upfront costs due to manufacturing were significant, they still represented a small percentage of the car's total lifetime emissions.

Finer-grained life-cycle analyses added some important caveats to that conclusion. One was that the type of lithium-ion cathode mattered greatly. Older lithium–cobalt oxide cathodes required twice as much energy to manufacture as newer lithium-ion nickel-manganese-cobalt cathodes, which were gaining popularity with automakers.[37] Life-cycle analyses also highlighted the advantages of scale: as battery production expanded, it became more efficient, which could reduce upfront environmental impacts.[38] And sourcing battery materials from recycled sources had the potential to

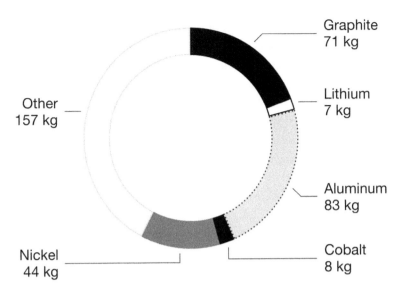

Graphite
71 kg

Lithium
7 kg

Other
157 kg

Aluminum
83 kg

Nickel
44 kg

Cobalt
8 kg

FIGURE 4.1 The type of lithium-ion battery employed has significant consequences for the weight of the battery, its material composition, and the places implicated in its production. The chart above represents the material composition of a nickel-cobalt-aluminum (NCA) battery pack with a 200-mile range. Data drawn from C. Xu et al., "Future Material Demand for Automotive Lithium-Based Batteries," *Communication Materials* 1 (December 2020).

significantly lower future manufacturing impacts. Contrary to Lomborg's assertions, from the perspective of reducing energy consumption and reducing greenhouse gases, electric cars could play an important role in phasing out fossil fuels. That was not to say there were not substantial upfront costs or that charging electric vehicles with coal-fired electricity made sense, but proponents looked at those costs as investments in the future. And the advantages of electric cars improved rapidly in the 2010s, as electrical grids around the world, including in the United States, began to shift rapidly away from coal, and the efficiency of battery manufacturing improved as the industry scaled.[39]

But even if life-cycle analyses penciled out in favor of electric cars, other questions complicated any simple assessments of their benefits. Was there enough lithium or cobalt to support a transition to electric cars? Might it be possible to phase out the use of cobalt in lithium-ion batteries? Would transitioning to electric vehicles mean a new dependency on imported batteries? Critiques such as Lomborg's and more complete life-cycle analyses contributed to a fuller discussion of the impacts of electric cars in the early 2010s. But such analyses were inherently reductionist, translating the complexities of vehicle manufacturing, use, and disposal into quantitative measures of energy consumption, greenhouse gas emissions, and other pollutants. Such analyses made possible broad comparisons, but they did little to wrestle with the specific ways in which electric cars intersected with the particular landscapes, communities, and people most affected by the extraction of the materials that made electric cars possible. And almost without exception, the places and people most involved in sourcing those materials were far removed from the showrooms where electric cars hit the road.

THE NICKEL CONNECTION

In 2016, Tesla's chief executive, Elon Musk, remarked that the term *lithium-ion battery* was a misnomer. "Technically, our cells should be called nickel-graphite, because the primary constituent in the cell as a whole is nickel."[40] Musk's statement was increasingly true for much of the electric car industry. To lower costs, improve range, and reduce the use of cobalt, electric car manufacturers deployed newer lithium-ion battery chemistries with increasing levels of nickel.

This marked a shift from the lithium–cobalt oxide chemistry that had driven the wireless revolution: the older cobalt-based chemistry that John Goodenough pioneered allowed only about half of the lithium ions to be checked out before the shelving system began to collapse. Nickel provided a sturdier shelving system, allowing almost all the lithium ions to migrate from the cathode to the anode as the battery charged, which meant a theoretical energy density almost twice that delivered by a lithium–cobalt oxide battery. And nickel had the advantage of being cheaper than cobalt and reduced the dependency on mines in the Democratic Republic of the Congo. In practice, however, pure lithium-nickel cathodes posed challenges too. Lithium-nickel cathodes were difficult to synthesize, prone to degradation, and thermally unstable.[41]

Solving those problems required doping the lithium-nickel cathode with other metals, such as aluminum, manganese, or even smaller amounts of cobalt, which improved stability, durability, and current flow. Bringing these cathodes into production happened in the early 2000s. By the early 2010s, lithium nickel-cobalt-aluminum oxide (NCA) and lithium nickel-cobalt-manganese oxide (NCM) had emerged as the dominant lithium-ion battery chemistries for electric cars. Tesla vehicles deployed NCA cathodes beginning with the Model S.[42] Most other automakers relied on variations on NCM, including Audi, BMW, Chevrolet, and Volkswagen. That pattern held true in China too, where nickel-based cathodes accounted for 90 percent of the electric car market by 2018.[43] But, to Musk's point, for every two miles of range, the battery required approximately one pound of nickel. That put a spotlight on the nickel industry.

Since World War II, the nickel industry had grown exponentially, driven by the demand for stainless steel. Nickel-based lithium-ion batteries promised a new market of the same scale. In 2017, electric cars accounted for 33,000 metric tons of nickel use. That was projected to grow to 570,000 metric tons by 2025, making it the fastest-growing market for nickel.[44] The question was whether there was enough nickel. Although nickel is the fifth-most-abundant element on Earth, and large bodies of nickel ore are known to exist around the globe, lithium-ion batteries required 99.8 percent pure Class I nickel. While most of the world's nickel is found in low-grade laterite ores, Class I nickel is most readily sourced from high-grade sulfide ores, production of which was dominated by Australia, Canada, and Russia in the early twenty-first century.[45]

Sourcing nickel from sulfide ores resembled other large-scale hardrock mining processes, such as those used in the lead industry: it depended on underground or open-cut mining, grinding ore, physical and chemical separation, and high-temperature smelting. Only a small fraction of the world's nickel output is refined into nickel sulfate, the brilliant blue-colored salt that serves as a precursor for a lithium-ion battery cathode. But historically, the leading supplier of nickel sulfate has been Norilsk Nickel, a Russian company with roots in the 1930s Gulag. It has led the world in both production of nickel ore since World War II and, more specifically, nickel sulfate in the twenty-first century. The early steps in its mining and smelting processes were centered in Norilsk, a Siberian mining town located nearly two hundred miles north of the Arctic Circle in central Russia.

The nickel mines around Norilsk tapped into the geological aftermath of what is known as the Siberian Traps eruption. Some 250 million years ago, a superplume of magma erupted in Siberia for several million years. The volcanic eruptions hurled massive clouds of dust into the air, loaded the atmosphere with greenhouse gases (likely by setting geological stores of coal and oil ablaze), and polluted it with volatile emissions.[46] To biologists, the Siberian Traps are best known for driving the Permian-Triassic extinction event, the largest mass extinction in the history of life. To metallurgists, the Siberian Traps are known for depositing plumes of magma from the Earth's core in sulfide veins rich in nickel, copper, platinum, and other minerals.[47]

Russians began developing the Norilsk mines in the 1930s. Political prisoners, forced into hard labor as part of the Russian Gulag, undertook much of the early work. In the 1950s, 68,000 of Norilsk's 77,000 residents were still convicts. After the Gulag system ended in the late 1950s, strong state support during the Soviet era, which included a generous "Nordic coefficient" that doubled the salaries of citizens willing to endure Norilsk's isolation, grueling work, and Arctic climate, helped expand the mining industry and the town.[48]

In 1970, Norilsk's Nadezhda smelting plant opened, making the Soviet Union the leading source of nickel globally. The city's population reached 135,000 that year.[49] The company faltered with the collapse of the Soviet Union in 1991 and the withdrawal of state subsidies before being sold at auction in the late 1990s to two Russian oligarchs, who aggressively expanded production in the 2000s. In 2010, Norilsk Nickel, renamed Nornickel, was one of Russia's most important industries, accounting for 2 percent of Russia's GDP and nearly 5 percent of Russian exports. It was also the country's single largest polluter, accounting for one-quarter of Russia's sulfur dioxide emissions.

Nickel extraction in and around Norilsk left behind a wasteland scorched by sulfur pollution. The mining itself largely took place underground. Miners working up to two thousand meters below surface extracted pentlandite, a nickel-iron-sulfide, that ran from 3 to 6 percent nickel, among other ores. The pollution stemmed from processing the ore above ground, where roasting the crushed ore liberated sulfur, which poured into the atmosphere as sulfur dioxide. In the early 2000s, before the Nadezhda Metallurgical Plant, its oldest smelter, was closed, Norilsk Nickel produced 120,000 tons of refined nickel per year and emitted 380,000 tons of sulfur dioxide from that factory alone—more sulfur dioxide than all of Germany that year.[50]

In 2006, Blacksmith Institute, an environmental watchdog organization, listed the entire Norilsk region as one of the ten worst-polluted sites in the world.[51] Acid deposition had scorched two thousand square miles of surrounding forests.[52] Air and water sampling revealed pollution levels exceeding Russian standards by more than an order of magnitude. Life expectancies for Norilsk employees were ten years below the Russian average and cancer rates nearly three times above average.[53] And, in 2008, its largest shareholder issued an open letter, urging the company to modernize its operations.[54]

"It is impossible to pollute for 80 years and then just stop very quickly," the director of Nornickel's Polar Division told the *Financial Times*.[55] In fact, there had been earlier efforts, including some post–Cold War era cooperative efforts to clean up Norilsk's operations under the auspices of the intergovernmental Arctic Council.[56] But starting in 2016, under pressure from Western investors, other Arctic countries suffering from wind-borne pollution originating at Norilsk, and, to some extent, Russian environmental authorities, Nornickel embarked on a massive program to replace or modernize its Soviet-era facilities.[57]

It closed its oldest Norilsk smelter in 2016. It began shipping concentrates from Norilsk to a newer refinery in Finland, which refined concentrates into various end products, including the nickel sulfate important to lithium-ion batteries. It began implementing improvements, including a sulfuric acid plant, at a cost of $2.5 billion with the goal of reducing sulfur dioxide emissions by 75 percent by 2023.[58] As part of these improvements, Nornickel even tied the city of Norilsk, home to 177,000 people, into the high-speed internet for the first time, stringing fiber optic cables six hundred miles across the Siberian tundra in 2019.[59]

In part, these efforts sought to reposition Nornickel at the forefront of a sustainable metal supply for a new generation of electric cars. That strategy paid dividends. In 2019, the German chemical giant BASF co-located its battery materials operations with Nornickel's Finnish operations, where it could rely on easy access to Norilsk's vast sources of nickel. In 2020, Elon Musk urged the expansion of global nickel production, promising a "giant contract" to a supplier who could supply nickel in an efficient and environmentally sensitive way—a contract Nornickel surely coveted.

Nornickel's efforts to improve its image suffered a black eye in May 2020, when a fuel tank collapsed near Norilsk, staining the Ambarnaya River red for six miles downstream and drawing international attention. Then, in August 2020, Aborigen Forum, a coalition of indigenous groups who had long objected to Norilsk's operations with little consequence, appealed directly to Elon Musk, Tesla's chief executive officer: "We are respectfully requesting that you DO NOT BUY" any metals from Nornickel until it provides a full assessment of its legacy of environmental damage in the Arctic, develops a plan for environmental restoration (including the 2020 oil spill), compensates indigenous groups for damage to "their traditional way of life," and acknowledges their standing under the United Nations Declaration on the Rights of Indigenous Peoples.[60]

FIGURE 4.2 Indigenous peoples around the Arctic protested the potential consequences of scaling up nickel production to meet Tesla's growing demand in 2020. Yana Tannagasheva was one of many who posted their protest pictures on social media. Courtesy of Yana Tannagasheva.

These were long-standing concerns for the Nenets, Enets, and Dolgan indigenous peoples whose homelands long predated the Norilsk metallurgical complex on the Taimyr peninsula. As a United Nations report documented, the mining operations had "irretrievably destroyed" the tundra reindeer pasture and many sacred sites important to their nomadic cultures. Despite Russian law requiring compensation to indigenous peoples for damage to traditional territories, Nornickel had offered little compensation as of 2013.[61] In 2018, however, in line with its efforts to improve the sustainability of its operations more generally, the company had adopted a policy that recognized the rights and autonomy of indigenous peoples, at least on paper.[62] Nornickel's efforts accelerated in the wake of the May 2020 oil spill and the Aborigen Forum's appeal to Musk to boycott Nornickel. In September 2020, the company entered into a five-year agreement to provide $25.6 million to fund housing, infrastructure, and healthcare, educational and cultural initiatives administered by local indigenous groups.[63]

In short, while companies like Tesla touted their efforts to reduce cobalt consumption in the 2010s and promised cobalt-free batteries in the future, longer-range and more durable lithium-ion batteries for cars often meant more nickel, which came with its own baggage. "Tesla will give you a giant contract for a long period of time if you mine nickel efficiently and in an environmentally sensitive way," promised Musk in 2020. "So, hopefully, this message goes out to all mining companies. Please get nickel."[64] While nickel producers with operations in Canada, Australia, and Russia all competed for Tesla's business, Norilsk's remained a leading producer of the high-quality Class I nickel suitable for batteries. As anticipation for an electric-car future ramped up, the biggest concern was that investments in nickel production were likely to lag demand, especially for battery-grade Class I nickel. Battery manufacturers could turn to lower-grade sources of nickel, such as the laterite deposits abundant in Indonesia and the Philippines, but that would require even more intensive materials processing, further complicating the goal of scaling up a sustainable electric car future.[65] Despite the strong interest in nickel-based lithium-ion battery chemistries, some automakers, including Tesla, hedged their bets and began to pursue alternative nickel-free lithium-ion batteries in the early 2020s too.

LITHIUM REDUX

Just as analysts expected, the surging demand for lithium-ion batteries, both for consumer products and electric cars, drove a massive increase in lithium demand in the 2010s—global production jumped more than sixfold.[66] But where that lithium came from was a surprise. Although the largest stores of lithium lay under the salt flats of South America's Lithium Triangle, where solar-powered extraction made low-cost production possible, the geography of the lithium industry shifted rapidly and unexpectedly between 2010 and 2020. The question was not just where the lithium was but how quickly the "white gold" could be brought to market.[67]

Most observers expected South America's Lithium Triangle to meet the growing demand. Multinational interests from China, Japan, South Korea, and the United States lined up to prospect for and develop salar-based lithium mines in Chile, Bolivia, and Argentina in the early 2010s. As the scholar Javier Barandiarán has explained, in the early 2010s, Bolivians, Argentinians, and Chileans saw in the global rush for lithium an opportunity to break the long-standing resource curse that had shackled their countries. Instead of just supplying lithium to the world, as they had historically supplied other raw materials, each country harbored hopes of developing a domestic manufacturing base for lithium-ion batteries.[68]

In Bolivia, site of the world's largest salt flats and lithium reserves, the Bolivian Congress and its socialist president, Evo Morales, refused to allow foreign interests to tap Bolivia's lithium resources in the 2010s, instead holding out hope of developing a domestic lithium-ion battery industry.[69] In Chile, a 2012 state-supported effort to expand SQM's mining operations on the Salar de Atacama ran aground over claims of corruption and illegality. In response, Chilean activists pushed the state to nationalize the lithium resource and renegotiate the generous contracts awarded to SQM and Chemetall, a position endorsed by a state-appointed national lithium commission in 2015.[70]

With new projects in Bolivia and Chile on hold in the 2010s, Argentina seemed best positioned to expand lithium extraction on its northern salt flats in Catamarca, Jujuy, and Salta provinces. Estimates put Argentina's potential lithium resources at 850,000 tons, which was more than thirty times global production in 2010.[71] Argentina already had one operating lithium mine—a brine operation on the Salar del Hombre Muerto started by the North Carolina–based company FMC in the mid-1990s—that made Argentina the world's fourth-largest lithium producer in 2010. That project had been developed in line with Argentina's export-oriented economic growth strategies in the late 1980s and early 1990s. But the investment climate was different in the early 2010s. Despite Argentina's historic support for foreign resource development, high inflation, rising corporate tax rates, and strict currency controls in the wake of the 2008 economic downturn unsettled the national investment climate.

But much of the challenge in mining lithium in Argentina was local. In the early 2000s, Argentina had slowly begun recognizing the communal landholdings of Argentine indigenous peoples. In 2009, thirty-three aboriginal communities protested early lithium prospecting, which they argued posed a threat to their land rights and had been authorized without proper consultation.[72] Their rallying cry was: "No comemos baterías, se llevan el agua se va la vida," which loosely translates to, "We cannot eat batteries, they take the water, life is gone." Across northwest Argentina, plans for lithium mining ran into strong local opposition, largely driven by indigenous groups, which had gained legal recognition and land rights. The other challenge that lithium prospectors faced was the geochemistry of the salars. Each salt flat

FIGURE 4.3 Indigenous communities have opposed lithium development in Argentina. This protest took place on February 15, 2019, in the Cauchari-Olaroz Salt Flats in the Salta province in northwest Argentina, blocking the main road to San Pedro de Atacama, Chile. Courtesy of Richard Bauer.

had a unique chemical signature, stemming from different concentrations of lithium, magnesium, and other impurities. That meant every new lithium brine operation required extensive sampling and pilot production runs before operations could be scaled up.

Between 2008 and 2018, the Argentinian salt flats saw a flurry of potential projects. Nearly fifty lithium operations—all run by multinational mining interests with some Argentinian investment—undertook at least preliminary prospecting. But securing the necessary permits (including from the provinces and indigenous groups), solving the site-specific chemistry of the lithium brines, and securing the capital needed to launch commercial operations deflated near-term prospects for Argentina's lithium boom.[73] By 2020, only one new project had reached commercial production. The Australian-Japanese project called Sales de Jujuy began shipping lithium carbonate to Toyota Tsusho Corporation in 2017. It touted its partnership with Olaroz Chico, the nearby indigenous community that permitted its operations. It provided jobs, supported local businesses, and made investments in local healthcare, schools, and infrastructure. Yet that project was the exception, not the rule.

In Argentina, Chile, and Bolivia, the pipeline from the salars to battery manufacturers proved much longer than anticipated.

While lithium production from the Lithium Triangle doubled between 2010 and 2018, Australian production grew sixfold, from 8,500 metric tons to 51,000 metric tons—pushing it past Chile and the rest of the Lithium Triangle to become the leading source of lithium globally.[74] Unlike the South American brine operations, which took years to bring online on account of local opposition, the slow rate of solar evaporation, and the peculiarities of brine chemistry, Australia mined lithium from hardrock. Hardrock mining operations could scale up production relatively quickly compared to the South American salars.[75] Australian lithium also shortened the supply chain, as it drew investment from and shipped lithium to China. While production costs in Australia remained higher, it also gained a key advantage in the 2010s. It was easier to manufacture the newer nickel-based lithium-ion cathodes important to electric vehicles from lithium hydroxide than from lithium carbonate.[76] Lithium hydroxide yielded a cathode with better structure, higher purity, and, ultimately, a longer cycle life.[77] Since it was easier to produce lithium hydroxide from hardrock than it was from brine, that gave the Australian manufactures an advantage. The downside was that it also increased the energy intensity of manufacturing a lithium-ion battery: producing lithium hydroxide from spodumene was substantially more energy and greenhouse gas intensive than sourcing lithium carbonate from brines.[78] But in the rush to secure more and more lithium, especially for a growing Chinese market, Australian-sourced lithium boomed in the 2010s.

In part, Australia succeeded in ramping up lithium production because lithium gave new life to old hardrock mines that had once been valuable for gold, iron, tantalum, or tin. Much like the Salar de Atacama in Chile, Western Australia had a history of mining that dated to the nineteenth century. What is today Australia's largest lithium mine, at Greenbushes in the very southwestern corner of Australia, began as a tin mine in 1888, making it one of the longest-operating mines in Western Australia. Lithium mining at Greenbushes began in 1983 and then expanded under the ownership of Sons of Gwalia, a gold, tin, and tantalum mining company, in the 1990s. After Sons of Gwalia went bankrupt in the early 2000s, Talison Minerals (now Talison Lithium) acquired the mine and made investments to double lithium production by the early 2010s.

Lithium extraction in Australia looked much like any other hardrock mining operation. Talison extracted spodumene pegmatite ore using open-pit mining, hauling ore to the surface in mining dump trucks, and processing it into concentrates through crushing, gravity separation, and flotation. Lithium makes up less than 2 percent of the Greenbushes high-grade pegmatite resources.[79] As a result, the most extensive aspect of the Greenbushes mine site are the piles of waste rock and tailings. These piles are contoured to mimic the local topography and to facilitate future ecological restoration. Altogether, the Greenbushes mine is four times the size of Central Park. In a region known for extensive mining activity, however, the scale and disruption at Greenbushes was far less consequential than nearby iron, coal, and bauxite mines that marred the landscape to the north. For instance, Australia's largest bauxite mine is five hundred times larger than the Greenbushes mine.[80]

In the 2010s, lithium began to be priced as if it were "white gold." Growing demand drove the price of lithium carbonate from under $5,000 per ton in 2010 to more than $15,000 per ton in 2018. As the expectations for electric vehicles grew and the price of lithium soared, so did investments in new mines in Australia. By 2018, in addition to expanded production at the Greenbushes mine in southwestern Australia, additional mines came online at Wodgina, Mount Cattlin, Mount Marion, and Pilgangoora. Two of the largest new mines to come online in the late 2010s were located in West Australia's Pilgangoora mining region. Unlike in Argentina, where indigenous groups often opposed mining operations, the Njamal indigenous group who held claim to the Pilgangoora mining region quickly reached agreements in exchange for employment opportunities and mining royalties to be paid into the Njamal People's Trust, which had been created in 2003 under Australian law to promote development and provide social services and financial support for the Njamal people.[81]

By 2018, one journalist crowed that the Australian mining industry had "stolen a march on the lithium world," seizing the lead in global lithium production and positioning itself as Asia's and, especially, China's chief lithium supplier. Although Australia only held a fraction of South America's lithium reserves, its history of hardrock mining, the ability to give new life to old tin or tantalum mines, and the relative agility of hardrock mining operations—compared to the yearlong process needed to produce lithium from brine—had allowed the industry to expand quickly.[82] Despite strong support for mining in Western Australia, lithium still remained an unknown to most banks and institutional investors. As a result, much of the investment in Australia's lithium sector had come from end buyers eager to secure a reliable supply of lithium. For example, in 2012, the China-based Chengdu

Tianqi Industry Group, which was Talison's largest customer, took a controlling stake in Talison Minerals that helped finance its expansion. And as Japanese and Korean buyers grew concerned about China's influence, they began investing in Australian operations too.

AN ELECTRIC CAR FUTURE, ALWAYS
JUST AROUND THE CORNER

Despite lithium miners' high hopes, the road to an electric car future remained bumpy. In the United States, despite billions of dollars in federal grants and loans to build a domestic supply chain for advanced batteries and electric cars and tax credits of up to $7,500 for buyers, only four hundred thousand plug-in vehicles were on the road in the United States by the end of 2015—well short of the Obama administration's goal of one million plug-in vehicles.

The Obama administration's efforts to jump-start the industry had proved troubled. Despite the Department of Energy's grants to battery manufacturer A123 and start-up car manufacturer Fisker Automotive, each company went bankrupt, in 2012 and 2013 respectively. Fisker contracted with A123 for batteries for its high-end electric cars, and Fisker was A123's largest contract, which drove A123's investment in the new Michigan battery factory. But delays in Fisker's production plans, an A123 battery recall, and lower-than-projected sales at Fisker undermined the viability of both companies.

For skeptics, the double failure of A123 and Fisker was just more evidence of the Obama administration's ill-conceived and expensive efforts to hasten a transition to a renewable energy future. By 2012, other recipients of federal grants, including solar panel manufacturers Solyndra and Abound Solar, energy storage company Beacon Power, and another battery start-up, Ener1, had also gone bankrupt. In the view of critics, especially conservative lawmakers, this was all evidence that big government should not be playing "venture capitalist" with taxpayers' money. Others alleged improprieties in how the loans were disbursed and managed.[83]

The majority of the loans funded by the American Recovery and Reinvestment Act had been productive. By 2016, the Department of Energy touted a stimulus loan portfolio of $36 billion. Despite the high-profile failures of Solyndra, A123, and Fisker, the program's overall default rate was less than 2 percent, similar to private-sector lending programs.[84] And in the case of the failed loans, the federal government ceased lending when the companies failed to meet benchmarks. For instance, in the case of Fisker, the automaker only received $192 million of its $529 million loan. Proponents of the Obama administration's efforts to support a clean energy economy pointed out that the advanced manufacturing loans to Ford, Nissan, and Tesla, which amounted to $8 billion, had supported thirty-six thousand jobs.[85]

Yet critics remained focused on the failures of the loan program, the millions of dollars wasted, and the government's failure to move even more swiftly to freeze the loans of failing companies and save taxpayer dollars.[86] As a result, in the 2010s, just as other countries, especially China, were accelerating investments in the clean energy sector, the United States reversed course. The storm of political controversy that swirled around the failed loans and conservative Republicans' growing clout in Congress and then the White House in 2016 largely short-circuited additional direct government support.

Of the recipients of stimulus-funded lending, Tesla emerged as the headliner for the Department of Energy's Advanced Technology Vehicles Manufacturing Loan Program. In January 2010, Tesla secured a $465 million stimulus loan, at an interest rate that was a fraction of that of a conventional loan, which provided capital to retool an old General Motors–Toyota factory in Fremont, California, to manufacture its new luxury sedan, the Model S.[87] At the time, Tesla's only product was a high-end sports car. Tesla began deliveries of the Model S in 2012, manufacturing the car in Fremont using battery cells made by Panasonic in Japan. By the start of 2016, Tesla had sold more than 120,000 electric cars, employed 13,000 people, and repaid the government-supported loan in 2013. By paying off the subsidized loan early, Tesla precluded the government from exercising stock options worth more than $300 million at the time.[88]

Despite the growing public attention swirling around Tesla, heightened by the 2014 announcement of the Nevada Gigafactory, a new luxury sports-utility vehicle in 2015, and rumblings about a new mass-market electric sedan, the soon-to-be-named Model 3, the market for electric cars and plug-in hybrids remained anemic in the United States. Sales of vehicles such as the Chevy Volt and Nissan Leaf were modest. Plug-ins accounted for only one in 150 new car sales in 2015, which was a decline from the previous year. Even boosters admitted that "cars with plugs are far from breaking into the mainstream."[89] And electric cars faced growing political headwinds, with the election of Donald Trump in November 2016. Trump championed his support for the fossil fuel industry, proposed eliminating federal subsidies for electric cars, and made clear his plans to weaken the ambitious fuel efficiency standards the Obama administration had finalized in 2012.[90]

In the mid-2010s, China emerged as the world's largest market for electric cars. The Chinese government saw electric cars as a solution to multiple challenges the country faced. They would reduce urban air pollution and dependency on imported oil. They were an easier sell in a country where long-distance travel was limited and most buyers were purchasing their first automobiles. And electric cars offered an opportunity to scale up a domestic automobile sector that could leverage China's expertise and manufacturing capacity in consumer electronics and advanced batteries. China had provided state support for advanced battery manufacturers since the early 2000s. Starting in 2013, it targeted new electric vehicle incentives at consumers, including upfront subsidies, reduced automobile taxes, and expedited paperwork. In 2015, China's subsidies lowered the average sticker price of a new electric car by 51 percent.[91] Sales surged, jumping from 1 percent to 5 percent of Chinese passenger vehicles sales between 2015 and 2018. While the United States puttered along toward an electric car future, China began to accelerate.

In the same years that Tesla drew outsized attention for its electric car ambitions in the United States, the Chinese manufacturer BYD emerged as the world's largest plug-in vehicle manufacturer. It was a transition that had been in the making since 2003, when BYD purchased a controlling stake in a small state-owned Chinese automaker to get a toehold in a growing industry.[92] BYD's near-term plan was to improve the automaker's efficiency, drawing on BYD's expertise in manufacturing, with a longer-term plan of leveraging BYD's battery expertise to get a head start on the market for electric cars. In 2008, BYD rolled out the first long-range plug-in hybrid electric vehicle, the F3DM, beating the Chevrolet Volt and the plug-in Toyota Prius to the market.[93] Although sales of the car numbered in the thousands and were limited to China, it put BYD in good position as China ramped up support for electric vehicles. By the time China announced plans in 2012 to put five million "new energy" vehicles on the road by 2020, BYD was already China's largest plug-in vehicle manufacturer, and the successor vehicle to the F3DM, the Qin, was China's best-selling plug-in vehicle. Unlike Tesla, which specialized in premium vehicles, BYD focused on selling affordable and utilitarian vehicles tailored for China's market, including passenger vehicles, buses, and taxis. As one observer put it, "BYD attracts a fraction of the attention" of Tesla, while powering "a transition electrified mobility that's moving faster in China than in any other country."[94]

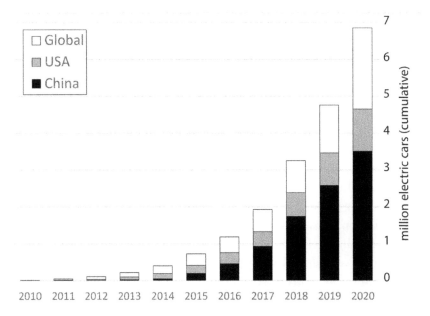

FIGURE 4.4 The stock of electric cars expanded quickly in the 2010s to reach seven million vehicles in 2020. Since 2015, China has accounted for the largest share of the world's electric vehicles. Courtesy of IEA (2021). Global EV Policy Explorer. www .iea.org/articles/global-ev-policy-explorer.

BYD's growth put it on the leading edge of China's rapidly growing electric vehicle industry. Well-funded start-ups in China, including Nio and XPeng, and foreign manufacturers, such as Tesla and VW, also expanded production in China (with foreign investment increasing after the Chinese government permitted direct foreign investment starting in 2018).[95] The growth in electric vehicle manufacturing in China was matched by equally large investments in battery production. In 2013, total automotive lithium-ion manufacturing capacity was less than 30 gigawatt-hours and located almost entirely in Asia. Despite the hullabaloo over Tesla's Nevada Giga-factory, which reached roughly 20 gigawatt-hours of production in 2018, that same year global lithium-ion battery manufacturing capacity reached 247 gigawatt-hours, of which 227 gigawatt-hours were located in Asia, and, of that, 148 gigawatt-hours were in China.[96] "Forget Tesla," pronounced one journal-ist.[97] That seemed a bit hasty, given that Tesla passed BYD to become the world's largest electric car manufacturer worldwide in 2019. But no matter how many cars Tesla or BYD produced, what was clear by the late 2010s was that each was almost entirely dependent on extensive supply chains that linked places like Australia, the Democratic Republic of the Congo, and Siberia to electric vehicles manufacturers by way of China.

In 2020, the United States remained almost entirely dependent on imported raw materials and chemicals for the primary components of a lithium-ion battery. Simon Moores, a well-known industry analyst, described factories like the Gigafactory more as "stand-alone achievements" than reflective of a "coherent US plan."[98] That stood in contrast to China, which had systematically developed capacity at each step of the lithium-ion battery supply chain. In 2019, although China accounted for only 23 percent of battery-related mining production, China dominated every other step of the expanding lithium-ion industry. According to Moores's consulting firm, Benchmark Mineral Intelligence, China accounted for 80 percent of chemical refining, 66 percent of cathode and anode production, and 73 percent of lithium-ion battery cell manufacturing. And, even in the case of mining production, although production was not located in China, Chinese companies, drawing on state-supported financing, were increasingly taking ownership shares in mining activities abroad.[99] Moores warned, "Those who control these critical raw materials and those who possess the manufacturing and processing know how, will hold the balance of industrial power in the twenty-first century auto and energy storage industries."[100]

In some places in China, that future already seemed to have arrived. By 2020, nearly all of the sixteen thousand buses and twenty-one thousand taxicabs servicing Shenzhen, China, the location of BYD's largest factory, were electric vehicles. Overall, despite China's beginning to scale back its direct subsidies for electric vehicle purchases in 2018, sales still topped 1.1 million in 2019, accounting for half of electric vehicle sales globally. In October 2020, China announced that all new cars sold after 2035 would be "new energy vehicles." Although the plans still allowed for plug-in hybrids, that made China's commitment by far the largest commitment to phasing out conventional vehicles.[101] In making that commitment, it was following the example of numerous European countries, Canada, and the state of California, which all aimed to phase out conventional vehicles, with deadlines set between 2025 and 2050. Some countries seemed well on their way to achieving those goals. All-electric vehicles sales climbed to 54 percent of car sales in Norway in 2020, as a result of aggressive policies that made electric vehicles cheaper to purchase and operate than conventional cars.[102] For the first time since Henry Ford put the Model T on the road in 1908, relegating the electric cars

FIGURE 4.5 Taxi drivers in Shenzhen, China, pass the time while their electric taxis charge. Note how large the charging station is and the integration of solar canopies. Courtesy of Billy H. C. Kwok.

to the shoulder of twentieth-century mobility, an electric car future once again seemed like it might be just around the corner.

CLOSING THE LOOP

A 2016 *Saturday Night Live* skit spoofed the Tesla Model S. It featured the actress Julia Louis-Dreyfus, of *Seinfeld* and *Veep* fame, touting a fictional Mercedes Benz AA class car as the "first fully electric luxury sedan powered entirely by AA batteries." With a straight face, Dreyfus described the car's "high-tech" drivetrain, "zero emissions," and a top-speed of 52 miles per hour. Better yet, she explained, the AA class did not need gasoline. It did not even require recharging. It ran on disposable AA batteries. When the batteries needed to be replaced, a special button ejected all 9,648 AA batteries from the vehicle, creating enormous piles of waste around the vehicle. The parody offered a sly commentary on an issue that cast a growing shadow over the growing enthusiasm for electric vehicles: what would happen to the batteries?[103]

Right from the start, electric car manufacturers hitched their future to the recycling bandwagon. Not only were electric vehicles going to be "zero emissions," but they were also going to be "zero waste." To put a shine on their new cars' green credentials, Nissan and Chevrolet touted the recycled materials that went into their new electric cars. They described seats and interiors partially manufactured from materials recovered from old plastic water bottles. And Nissan, Tesla, and Chevrolet all promised they were planning for the end of life of their vehicles even before the new cars hit the road. Nissan promised that 99 percent of the new Leaf was recyclable and that the lithium-ion battery was "fully recyclable."[104] In 2008, Tesla explained it had partnered with a California-based recycler to handle retired battery packs from the Roadster, recovering almost 70 percent of the valuable metals and oxides in the battery.[105]

In retrospect, such promises were wildly optimistic. Despite the billions of lithium-ion batteries that had already been deployed to power laptops, cell phones, and other consumer products since the early 1990s, lithium-ion battery recycling remained largely nonexistent in 2010—it lagged far behind lead-acid battery recycling, which remained the "gold standard," and even the limited programs for recycling disposable AA and AAA batteries. In large part, this was because lithium-ion batteries were hard to recycle and posed no immediate threat to public health. For instance, the process Tesla outlined involved deep freezing the battery in liquid nitrogen before shredding it. Unlike lead-acid batteries, which were largely standardized, highly toxic, relied on a largely uniform chemistry, and were profitable to recycle, none of that held true in the case of lithium-ion batteries.[106]

But, even if lithium-ion battery recycling was underdeveloped in 2010, high volumes of large-format electric car batteries promised to drive the economies of scale necessary to create the recycling infrastructure needed to make electric cars "zero waste." Analyses made clear that recycling was an imperative. In the near term, end-of-life electric vehicle batteries posed a fire risk if not decommissioned carefully. They could not easily be stockpiled or simply disposed of in bulk. In the longer term, closing the circle on electric vehicles batteries was going to be important to curb the demand for raw materials needed for batteries. In some scenarios, recycling had the potential to reduce the cumulative demand for the lithium, cobalt, and nickel needed to electrify the transportation sector by up to 30 percent between 2020 and 2050.[107]

Despite the urgency of large-scale lithium-ion battery recycling in the 2010s, developing such systems involved several key uncertainties. One question was when electric car batteries would be available for recycling. Engineers estimated the service life of electric car batteries at five to twenty years. But that estimate was complicated by another strategy for managing old batteries: the potential for "second life" applications. Although electric vehicle batteries would likely be retired when they lost 20 or 30 percent of their usable capacity, they could then be repurposed for stationary applications, such as providing battery backup for homes and businesses and helping meet periods of peak demand for electrical grids.

Right from the start, automakers and engineers gave substantial attention to second-life applications. GM piloted such a redeployment in 2012, repackaging five Volt batteries into a system that could provide backup power to three to five average American homes for two hours.[108] Nissan piloted a program to repackage Leaf batteries to offset peak power demand at car-charging stations. In 2016, Daimler harnessed one thousand retired battery systems from its micro Smart cars to provide 13 megawatt-hours of electricity storage to support renewable energy deployment on Germany's electrical grid.[109] Analyses of such strategies indicated that even after factoring in the need to transport, refurbish, and deploy such systems, they yielded significant environmental benefits. But such uses also extended the useful lifetimes of lithium-ion batteries significantly, which created uncertainty for recyclers.

In addition to when batteries would be available in volume for recycling, another uncertainty was the mix of lithium-ion batteries that would be available for recycling. Although the lithium–cobalt oxide cathode had dominated portable applications since 1991, the use of nickel-based cathodes had increased since 2010, especially for large-format batteries in electric cars. That affected the particulars of how lithium-ion batteries were handled. Did they need to be sorted? Did different chemistries require different recycling processes? And how valuable were the recycled materials? In the 2010s, to the extent that lithium-ion batteries were recycled, high-value cobalt drove the economics of recycling. Most observers anticipated that the declining use of cobalt would make it more difficult to finance lithium-ion battery recycling.[110]

Recycling lithium-ion batteries also raised questions about environmental sustainability. Although it was clear that recycling would reduce demand for newly mined raw materials and, as a result, pollutants such as sulfur dioxide that were closely tied with primary metals production, recycling still required significant investments of energy and resources. Traditional pyrometallurgical recycling processes that operated at high temperatures were especially energy intensive, wasted lithium, and recovered only a fraction of other metals, such as copper, nickel, and cobalt. Hydrometallurgical recycling processes recovered most of the metals, including lithium, but required intensive use of chemical reagents, water, and, in some cases, energy to operate processes at high temperatures and pressure. And neither pyro-metallurgical or hydrometallurgical processes fully closed the loop on batteries: for instance, neither recovered graphite or the electrolyte, which together accounted for more than half of a battery's mass.[111]

Some of the most promising strategies employed direct recycling, which instead of reducing batteries to their underlying raw materials, separated the cathode and anode materials, retaining their basic properties—such as the cathode's crystal structure—and then chemically reconditioned them to restore their electrochemical properties—such as replenishing lithium lost to degradation or during recycling. In general, such processes required fewer inputs and produced less waste. The viability of such strategies, which remained in the pilot stage in 2021, was that they depended both on the electrochemical state of the retired battery—a fully discharged battery was harder to process than a charged one—and had to be carefully tailored to match the particular cathode chemistry. The upside was that pilot projects indicated that such strategies might begin to close the loop on lithium-ion batteries with minimal environmental impacts, by directly recovering materials from the cathode and anode for reuse.[112]

In short, contrary to the strides in manufacturing and deploying lithium-ion batteries at scale, there was still much work to do to develop the recycling processes and infrastructure required to live up to the promise that electric car batteries were "fully recyclable." That possibility attracted an intense level of research at universities, government research laboratories, and companies in the 2010s, especially in China, as the scale of spent lithium-ion batteries grew.[113] Ultimately, achieving that goal will require more than just improvements in downstream recycling processes; it will also depend upon upstream changes in how batteries are manufactured and handled: prohibiting the disposal of electric vehicles batteries; requiring automakers to disclose the material composition of battery packs; incentivizing battery manufacturers to use water-soluble binders to facilitate separating the cathode, separator, and anode at end of life; and making manufacturers legally responsible for end-of-life processing of electric vehicle batteries.

While countries adopted aggressive policies to ramp up the use of electric vehicles in the 2010s, they played catch-up on recycling policies. In the United

States, several states including California prohibited disposal of electric vehicle batteries, but lithium-ion batteries were not classified as a hazardous waste under existing federal regulations and therefore were unregulated nationally. Although some states, such as California, worked on regulations specific to electric vehicles batteries in the late 2010s, and the Department of Energy launched a lithium-ion battery recycling research center headquartered at Argonne National Laboratory, much of the activity was centered in the private sector. Despite Tesla's vision of integrating recycling directly into the Gigafactory's operations, in 2017 one of its cofounders launched a new enterprise, Redwood Materials. It aimed to develop a large-scale lithium-ion recycling battery operation that could "short-circuit" the existing mining industry with a low-cost, large-scale domestic supply of raw materials that would turn "waste into profit" and help "save our planet." In September 2021, Redwood Materials announced plans to build a battery materials manufacturing factory in North America to support production of one million electric car batteries by 2025 and five million electric car batteries by 2030.[114]

Both the European Union and China actively developed recycling policies in the late 2010s. The European Union classified electric car batteries as industrial batteries under existing 2006 regulations, prohibiting disposal, requiring manufacturers to fund collection and recycling, and mandating a minimum recycling efficiency of 50 percent.[115] It was in the process of updating those regulations in 2020 to ensure that recycling delivers "a significant share of the raw materials required for battery production" in the European Union.[116] In 2018, China issued policies that made manufacturers responsible for funding collection and recycling of batteries; required that advanced recycling processes consider recycling efficiency, safety, personnel, and environmental protection; provided for tracking the lifetime of individual batteries; and encouraged the development of design standards to facilitate reuse and recycling of batteries.[117] Similarly to those of the European Union, China's policies aimed to formalize an extended producer responsibility system to improve recycling and curb waste by making producers financially responsible for their products at end of life.

The volume of spent lithium-ion batteries from electric vehicle batteries is projected to surge in coming years, surpassing seven hundred thousand tons annually by 2025 (that same year an additional four million tons of batteries are projected to be put on the market). Historically, estimated recycling rates for lithium-ion batteries have been low, hovering between 5 and 10 percent. Since 2019, however, new research suggests that lithium-ion battery recycling has scaled up more rapidly than understood, primarily in China and South-east Asia, and approaches 50 percent.[118] This is promising news: it suggests that as recycling costs fall, recovery efficiency improves, and the price of raw materials increase, lithium-ion battery recycling is becoming viable—the concentration of recycling in Asia is closely tied to local demand from battery manufacturers. But, if the goal is not just to close the loop on lithium-ion batteries but also to diversify the geography of lithium-ion battery recovery, this should also be read as a warning sign. Although there are growing private-sector and governmental initiatives to promote recycling in the United States, the flow of spent lithium-ion batteries, and the resources they contain, could easily be diverted overseas. Lithium-ion battery recycling is a risky enterprise, made uncertain by changes in technology, industrial processes, and government policy. Policy action, investments in infrastructure, and harmonization developed in the near term in the United States and else-where will play a pivotal role in how sustainably the world can manage the surge of retired lithium-ion batteries coming in the 2020s, and how well distributed the recovery of those resources will be.

In the mid-1990s, the United States Advanced Battery Consortium set a goal of reducing battery costs to below $150 per kilowatt-hour of capacity. In their assessment, that was the threshold at which electric vehicles would become cost competitive with conventional cars.[119] In 2010, battery pack costs still hovered around $1,100 per kilowatt-hour. At the time, even optimistic projections anticipated battery costs would remain at more than $250 per kilowatt-hour in 2020.[120]

Yet lithium-ion batteries outpaced expectations in the 2010s. Battery prices plummeted, falling to $110 per kilowatt-hour in 2020.[121] The scale of production rocketed past projections of 120 gigawatt-hours to reach 345 gigawatt-hours annually. By 2020, analysts began projecting that the purchase price of an electric car would soon reach cost parity with conventional vehicles, even without subsidies. As one observer commented, "If you asked anyone in 2010 whether we would have price parity by 2025, they would have said that was impossible."[122]

No single breakthrough in lithium-ion battery technology drove electric cars toward price parity. Instead, lithium-ion batteries advanced in increments. Production and deployment expanded progressively, ramping up from smaller- to larger-scale and shorter- to longer-lived applications: cell phones in the 1990s, laptops and smartphones in the 2000s, and electric cars in the 2010s, which laid the groundwork for grid-scale lithium-ion battery storage applications—such as providing storage for renewable energy systems and meeting peak power demand—in the 2020s. Over thirty years, as each rung of this ladder was erected, from smaller- to larger-scale applications, global supply chains developed, manufacturing capacity expanded, technical know-how advanced, and costs fell. This implementation ladder made it possible to "bottle lightning" at larger and more useful scales, putting a clean energy future within reach.[123]

In recent years, the electric car industry has seemed remarkably nimble. China brought new battery megafactories into production at the rate of one per week in 2019. Tesla began shipping new electric sedans from its sprawling Shanghai Gigafactory less than a year after breaking ground in December 2019. But further up the supply chain, the story of the 2010s was different. Bringing new mines and chemical processing facilities into production moved much more slowly, often plagued by production challenges, slowed by local opposition, and complicated by lengthy permitting requirements. For instance, annual lithium-ion battery sales grew tenfold in the 2010s; the supply of lithium grew one-third as fast. Despite the burst of new lithium capacity in Australia, new supplies of lithium, nickel, graphite, and cobalt lagged behind. As Benchmark Minerals reported in 2020, the rapid decline in the price of lithium-ion batteries historically had been driven by two things: "economies of scale and technological improvements." To think that will continue, Benchmark Minerals warned, ignores "the cost of raw materials."[124]

In short, the emerging bottleneck in the lithium-ion battery industry was not at the final stages of assembly, which hinged on the precise work of combining the already highly refined chemicals that made up the cathodes and anodes into battery cells, packs, and eventually electric cars. The bottleneck was sourcing and synthesizing those materials in the first place. The materials for lithium-ion batteries are as much made as they are mined; building the mines and capacity to refine ore into battery-grade materials was capital- and energy-intensive work that generated waste by the ton and depended upon processes where impurities on the order of parts per million mattered. While it might take one to two years to start up a new battery cell factory, industry analysts estimated it took three to five years to bring a chemical-processing facility online and five to twenty-five years to bring a new mine online. And looking at the pipeline of new mining projects and chemical-processing facilities globally, industry analysts saw a widening gap between the demand for lithium-ion batteries and the supply of materials. No doubt higher prices and growing markets will spur production, but building mines, it turns out, is not like building battery megafactories: it takes more time. Tesla's Nevada Gigafactory was a case in point. It proved much easier for Tesla to ramp up battery cell production—which largely happened on schedule—than it did to develop North American supplies of lithium, graphite, and other battery-grade materials—which lagged far behind its goal of domestic self-sufficiency.

Amid the excitement over the falling price of batteries, aggressive goals for electric vehicle adoption (Norway's ban on conventional vehicles is set for 2025; President Biden set a goal of making 50 percent of car sales in the United States electric by 2030), and high-profile events like the unveiling of Ford's F-150 electric pickup truck in 2021, it is still easy to lose sight of what is at stake. For the challenge is not just ramping up manufacturing, securing supplies of materials, and transitioning to a clean energy future. The challenge is doing so sustainably and fairly.

While the benefits of a clean energy transition will be global, the short history of lithium-ion batteries makes clear that the impacts of materials production that make them possible will be borne by local communities: the indigenous peoples living on the periphery of Andean salars, the residents and workers of the Katanga region of the Democratic Republic of the Congo, the indigenous peoples and residents of the Norilsk region of Siberia, or the rural farmers neighboring mining sites in China. Threads of these communities' stories of losses and burdens are woven into the cathodes and anodes of nearly every lithium-ion battery circulating in the world.

This offers a sharp reminder that a "just transition" cannot succeed only by fostering "energy democracy" and a "regenerative economy"—concepts that invest in the local stewardship of energy production and resource management. Instead, it must reform the practices of extraction too, and the global trade in materials that will be essential to remaking energy infrastructure in communities and countries worldwide. That will mean supporting new mining operations, trade policies, and certification schemes aimed at ensuring that the coming boom in materials production at home and abroad does not reproduce the injustices of the fossil fuel past. That poses a challenge and an opportunity that is essential to accelerating the transition to a clean energy future.

BUILDING A CLEAN ENERGY FUTURE FROM THE GROUND UP

A FEW DAYS BEFORE EARTH DAY 2019, ALEXANDRIA OCASIO-CORTEZ, A member of the House of Representatives from New York, released a "message from the future." It was a seven-minute animated video featuring Representative Ocasio-Cortez looking back at the 2020s as the pivotal moment in American history when the United States met the challenge of climate change.

The first minute of the video held few surprises. It laid out the climate problem in the usual way: corporations such as Exxon had ignored the warnings of scientists, including their own, and "kept digging and mining, drilling, and fracking like there was no tomorrow." As a result, the world needed to halve greenhouse gas emissions by 2030 according to climate scientists. Otherwise, Ocasio-Cortez warned, the future would be full of more "climate bombs," like Hurricane Maria, which had devastated Puerto Rico in September 2017.

What came next in the video reflected a key turning point. For three decades, those concerned about climate change had spoken with far more conviction about what they opposed—fossil fuels, deforestation, big agriculture, and greedy corporations—than what they were for. But Ocasio-Cortez's video reflected the growing enthusiasm for a Green New Deal: a proposal for systemic change to tackle not just climate change but the inequities in American society.

At the heart of the Green New Deal was a "swing-for-the-fence" strategy that aimed to leverage the climate crisis to transition the nation to 100 percent clean energy, transform the economy, create jobs, provide healthcare, and promote social equity, including addressing the historic oppression of "frontline and vulnerable communities." To observers, it may have seemed "Too big!" or "Not practical!" or "Too Fast!" In response, Ocasio-Cortez made clear, "as a nation, we had been in peril before."[1]

The Green New Deal called for "a new national, social, industrial and economic mobilization on a scale not seen since World War II and the New Deal era."[2] Observers enthused that by setting aside a narrow focus on controlling pollution or pricing carbon and engaging in issues of economics, equity, and justice, the Green New Deal demonstrated what "realistic environmental policy looks like" for the 2020s.[3] Even if prospects for enacting the Green New Deal appear slim, considering the partisan divides in American politics, the proposal offers an important reference point for the types of policies, the scale of change, and the conceptual scope of the plans needed to tackle climate change.

What lessons does the history of batteries offer on the cusp of a clean energy transition at the scale envisioned in the Green New Deal? Too often, the material consequences of scaling up a clean energy future, and the implications that shift might have for matters of social and environmental justice, fall to the side amid the increasingly urgent calls to wean the world off fossil fuels and slow the rate of climate change. Considering the history of batteries and prospects for a clean energy future makes plain the urgent need for better policies that regulate the extraction, production, and trade in materials and goods across borders, especially as we envision a more sustainable future that can only be built by massively scaling up the supply of raw materials, metals, and chemicals—increasingly, the very stuff of modern life.

THE MATERIAL IMPLICATIONS OF A GREEN NEW DEAL

Naomi Klein's *This Changes Everything: Capitalism vs. the Climate* is on the short list of books that inspired the surge of activism around climate change and social justice in the 2010s. It wove together growing concerns about climate change and corporate power and made them newly relevant in the aftermath of the Great Recession and on the leading edge of the climate crisis. The centerpiece of Klein's argument was a forceful critique of what she labeled "extractivism." In her analysis, extractivism was a mind-set, a ruthless agent of capital accumulation and social inequity, and an enormously destructive form of resource production that drove the fossil fuel industries and climate change.[4]

If Klein's *This Changes Everything* offered the most compelling case against the perils of a fossil fuel–powered capitalist economy, then the Green New Deal can be read as a blueprint for change on the scale, at the pace, and with the scope necessary to address climate change and rein in capitalism.[5] For progressive climate activists, it was not enough to transition to a net-zero economy by 2050; the imperative was a just transition. Klein emphasized what a clean energy transition would *create*: energy independence, community resilience, millions of desirable jobs, and, ultimately, an alternative to fossil fuels.

What went largely unspoken by Klein, proponents of the Green New Deal, and environmentalists more generally was what a transition to a clean energy future would *consume*. Deploying solar panels across rooftops, wind farms across square miles, and electric vehicles by the millions will drive a dramatic increase in demand for energy-relevant minerals, with implications for frontline communities around the world. The minerals needed to enable a clean energy future are projected to approach two hundred million tons per year by 2050.[6]

Mobilizing materials on this scale poses enormous challenges, but it is far preferable to the fossil fuel status quo. In 2019 alone, the global coal industry produced eight thousand million tons of coal. Even when factoring in the waste associated with processing ores into the high-quality metals needed for clean energy technologies (which ranges from a factor of four for aluminum production up to a factor of one hundred or more for cobalt), the materials throughput to support a clean energy transition is still roughly half the total material requirement of a business-as-usual scenario that relies on fossil fuels.[7] That's the good news. But mobilizing materials at this scale, without reproducing the harms and inequities of the fossil fuel–powered twentieth century, poses an urgent challenge.

THE IMMATERIALITY OF MODERN ENVIRONMENTALISM

If the material implications of transitioning to a clean energy future loom so large, why have environmentalists, climate activists, and progressive thinkers been slow to make the material implications of a clean energy transition central to their proposals? Why have the Green New Deal and similar proposals not included regulatory reforms to expedite new mines, goals for expanding raw materials processing, or strategies to promote fair trade for raw materials?

For all that was visionary about the Green New Deal, especially marrying social justice to a clean energy agenda, its aversion to wrestling with the material implications of a clean energy transition is rooted in familiar currents of antimodernism that run deep in environmental thought: an opposition to extractive industries, suspicion of synthetic materials, and discomfort with scale.

This first current of antimodernism is one that I know well. My first book is a history of the campaigns that environmentalists mounted to protect wilderness in the United States.[8] After the passage of the Wilderness Act in 1964, environmentalists often measured success in the millions of acres of public lands permanently protected from logging, oil and gas development, and mining in perpetuity. In the details, the wilderness movement has a rich history with surprising ties to social reform, labor history, and even indigenous communities, especially in Alaska. But, writ large, the wilderness movement often traded in distinctions between what was good (preservation and wildness) and what was bad (extraction and development), which translated into the knee-jerk opposition to extractive industries characteristic of modern environmentalism. It was such opposition that Klein so eloquently linked to issues of social justice in *This Changes Everything*.

Environmentalists have also long harbored concerns about "synthetic materials." Rachel Carson famously raised this concern about pesticides, such as DDT, in *Silent Spring* in 1962. Barry Commoner, a prominent ecologist and environmental thinker, focused on the threats that stemmed from the rapid pace of technological innovation and industrialization in his 1971 book *The Closing Circle*. This is the second current of antimodernism. Although Commoner never singled out batteries, he did point to mercury, lead, nickel, cadmium, and sulfur oxides as representative of the problems of modern technology and industrialization. The brute force with which such materials were extracted from the Earth, the high temperatures needed to refine them, and the resulting pollution all exemplified how "human acts have broken the circle of life." Commoner urged a transition away from "ecologically costly synthetics" and toward "goods produced from natural products."[9] Commoner was not alone among environmental thinkers in urging a return to "material simplicity."[10]

These concerns about extraction and synthetics dovetailed with a third current of antimodernism, which was environmentalists' discomfort with large-scale human activity—whether industrial, technological, or economic. The "appropriate technology" movement in the 1970s aimed to resolve the tensions between technological innovation and environmental protection through smaller-scale, human-centered systems that could strike a new balance between humans and the environment, valuing nature and human labor in ways that recognized the dignity of both. Its proponents envisioned a future of small-scale communities powered by windmills and solar power, grounded in material and energy efficiency, and supported by local agriculture and small-scale manufacturing. The "appropriate technology" movement promised not only to lighten the impact on the planet but also to promote decentralization and democracy. In the words of a leading proponent, Ernst Schumacher, it promoted enterprise and technology at scales that were "natural, fruitful, and just."[11]

Implicit in these concerns—the consequences of resource extraction, the hazards of synthetic technologies, and the scale of modern development—is the assumption that the global community, led by wealthy consumers in the developed world, must substantially reduce the consumption of goods, resources, and, ultimately, energy, to realign modern society with the limits of the natural world. In the past decade, some environmental thinkers have refined these ideas into a vision of "sustainable degrowth." In place of a sustainable growth paradigm, predicated on economic expansion and technological advance, sustainable degrowth hinges on reducing economic activity, downscaling consumption, and giving up on the "fantasy" of decoupling economic growth from environmental impacts.[12] In short, taken together, these tenets of modern environmentalism challenge the very bases of modern society, especially in more-developed countries such as the United States. Indeed, for some environmentalists, that is the point.

Despite the cogency of that charge, it is not clear to me that this constellation of ideas is up to the task of guiding us to a more just and sustainable future, given the scale, the urgency, and the deep inequities inherent in the climate crisis. Right now, more than anything, policies to accelerate a clean energy transition and address climate change require a broad base of public support. The politics of degrowth asks much of those who are already energy secure (of whom I am one), and it offers little to the majority of people in the world, both in the United States and abroad, who are energy insecure. For this latter group, the most urgent concern is not limiting energy consumption but expanding it to provide for basic needs.

This helps explain what has made the Green New Deal such a powerful and expansive vision for climate action. Instead of placing the burden on individuals and focusing on consumerism—as the politics of environmentalism often does—the Green New Deal instead offers a positive vision for significantly expanding the role of government in driving a clean energy transition that avoids the worst consequences of climate change while creating jobs, promoting equity, and addressing systemic injustices by meeting basic needs for healthcare, housing, economic security, and a healthy environment. This is not to dismiss the enormous challenges in effecting such reforms, which will challenge corporate power and likely require raising taxes on the wealthy. But in breaking with the usual austerity politics of environmentalism, the Green New Deal offers the promise of a broad-based environmentalism that links a clean energy transition to improving public welfare, especially for those most in need.[13] But what proponents of the Green New Deal have given less attention to are the implications of making this transition for the workers, communities, and environments that will be on the front lines of the coming boom in extraction needed to build this sustainable future.

TOWARD A MATERIAL ENVIRONMENTALISM

Just as addressing climate change demands careful forethought and aggressive action, with clear goals for ramping up renewable energy sources, deploying electric vehicles, and encouraging behavioral change, so too will building the mines, the supply chains, and the recycling infrastructure needed to enable a more sustainable and just clean energy future. Before narrowing in on the practical work of enacting policies needed to support a clean energy transition, it is worth considering some broader lessons about the relationship between materials, sustainability, and the environment that stem from the environmental history of batteries. Although these ideas have largely been on the margins of modern environmental thought, taken together, they offer useful waypoints as we chart a path toward what I described in the introduction to this book as a form of industrial ecological literacy.

One lesson is about prospects for "dematerialization." Starting in the early 1990s, ecological optimists began surveying the sweep of modern history. Unlike the degrowth advocates, who raised concerns about industrialization and advocated for a return to material simplicity, a new group of thinkers saw in science and technology the forces of "liberation" that could "decouple" society from "demands on planetary resources." Rockefeller University professor Jesse Ausubel, along with others, formalized the concept of "dematerialization," defining it as "the decline over time in the weight of materials used to meet a given economic function." The transitions from steel to aluminum, wood to plastics, petroleum to natural gas, and even prospects for the digital office all served as examples of dematerialization. As Ausubel and his colleagues read it, "the message from history is that technology, wisely used, can spare the earth."[14]

The familiar email tagline—"Please consider the environment before printing"—makes clear that trends toward dematerialization were never so simple. But, in the case of batteries, the theory largely holds: a lithium-ion battery stores almost four times as much energy per unit volume and five times as much energy per unit weight as a comparable lead-acid battery. And, more broadly, the trend toward electrification advances the case for dematerialization: although an electric vehicle requires a wide range of materials, overall, it is less material intensive than a conventional vehicle fueled by gasoline and spewing carbon dioxide from its tailpipe.

Even if there is a broad trend toward dematerialization, however, what the history of batteries makes clear is that the details of that trend matter. Despite environmentalists' hopes for "material simplicity" in the 1970s, the trend toward advanced technologies has driven another change: an increasing diversity of materials use. One of the little-appreciated trends of modern environmental history is how humans have harnessed the desirable properties of an expanding array of highly refined metals and chemicals, ranging from more-publicized rare earth elements to lesser-known materials, such as graphite. This has contributed to what researchers have described as the "omnivorous diet of modern technology": functional improvements in many modern technologies are dependent on more and more specialized materials that are drawn from across the periodic table.[15]

This is a second lesson that the history of batteries illuminates: how resource omnivory has become characteristic of technological advance. The lowly beer can was once made of steel; now it is fabricated from three different aluminum alloys, each optimized for the body, top, or tab.[16] A microprocessor relied on twelve elements in the 1980s; its elemental diversity exploded to more than sixty elements by the 2010s.[17] Since 2000, the simple incandescent lightbulb has been replaced by far more efficient, but materially complex, LED bulbs. The pattern holds in the case of batteries, too.

Compared to lead-acid batteries or single-use zinc-carbon batteries, lithium-ion batteries depend upon a growing array of cathodes, using a greater variety of elements, which deliver batteries optimized for power, capacity, durability, and safety.

What is valuable to material scientists, however, is not just a broad palette of materials but also the availability of high-purity materials that can serve as precursors for precision-engineered compounds. Battery-grade materials include refined forms of graphite, nickel, cobalt, and lithium, for instance. Such high-quality materials are not simply mined; they are manufactured. From early lead-acid batteries to the most recent lithium-ion batteries, the ability to deliver precisely engineered materials with minimal impurities and specific levels of desired dopants has been crucial to improving battery performance. Producing these battery-grade materials has often followed what I describe as the "paradox of purity." The higher the degree of purity, the more processing and the more energy the processing entails. This paradox only intensifies as lower-grade ore bodies are brought into production, as demand exhausts higher-grade ores. Thus, this trend toward "resource omnivory" comes paired with the "paradox of purity"—chalk that up as a third lesson.

These ideas—dematerialization, resource omnivory, and the paradox of purity—help me, as an environmentalist, begin to wrap my mind around the complex ways in which modern technologies entangle me and modern society as a whole in places and processes that span the globe. It is my contention that, if we are going to advance a more sustainable future, we need to think just as carefully about the stuff we consume as we do the places we recreate, the things we eat, or the water we drink. This is not a new idea, but it is one that is still hard to act on. Taking this approach recasts the significance of batteries, along with other modern technologies. Instead of seeing batteries as simply a technology or a consumer product—descriptions that distance them from the realm of nature and elevate them as evidence of human's "mastery of nature"—we need to read batteries as yet one more technology that is entangling humans in a world that is neither technological nor natural but a complex hybrid of the two. That points toward a fourth lesson: instead of pining for a return to nature or a world of material simplicity, we need to take responsibility for our deepening and inescapable "entanglement" in the material world and the cascade of consequences that has for both human and environmental health.[18]

Taken together, the ideas important to a more material environmentalism point toward the need for climate activists—who are concerned both about addressing climate change and ensuring a more just and equitable future—to scale up efforts to develop sustainability-minded resource policies and support investments in manufacturing, mining, and trade commensurate with the material challenge of a clean energy future. Four broad priorities should inform such policy initiatives:

Leverage Government Leadership to Build a Clean Energy Future from the Ground Up

The importance of government leadership may seem obvious to many proponents of a clean energy future, but it is useful to set this point in a comparative context. I know that I underestimated the scale and significance of China's state-sponsored investments in clean energy manufacturing and materials sourcing during the 2010s. How important could such investments be if China was still building a coal-fired power plant every week on average?

It turns out, very important. Starting in the 2000s and accelerating in the 2010s, China leveraged government support to align higher education, research institutes, private enterprise, and governments at all levels behind a strategy that elevated batteries and electric vehicles as core components of a national strategy that aimed to promote industrial development, improve energy security, and reduce air pollution. By 2018, China had become the world's largest producer and consumer of electric vehicles, with sales of more than one million vehicles. That head start in batteries for electric vehicles gives China an advantage in deploying grid-scale battery storage projects too.[19]

While US policymakers bickered over the collapse of Solyndra, A123, and Fisker—all beneficiaries of the Obama administration's clean energy investments—China took a different path. China, building on its already well-established consumer-electronics manufacturing sector, accelerated investments in battery and electric vehicles manufacturing, setting ambitious goals for electric vehicle sales in 2015 and 2020. To support those goals, China expanded subsidies to encourage electric vehicle purchases, leveraged municipal purchasing power, exempted electric vehicles from most taxes and vehicle-use restrictions (such as those imposed to limit pollution), and encouraged public investment in an extensive charging network.

Lesser known, however, is the extent to which China invested in the supply chains important to supporting its clean energy policies. Subsidies for electric car purchases favored Chinese-assembled vehicles using Chinese-manufactured batteries. Chinese lenders, supported by the state government, aggressively financed Chinese investments in mines and processing facilities in China and abroad for key materials, including lithium, cobalt, and graphite. Those investments gave China an outsized role in the lithium-ion battery supply chains that started at mines around the world and usually ended at Chinese factories.

The lesson here is that if the United States has any hope of catching up to China and realizing the potential for a clean energy transition to create jobs, improve energy security, and advance sustainability—not just for electric cars, but solar panels, wind turbines, and the other technologies important to a clean energy transition—it is going to require a massive mobilization of government support that includes not only prioritizing manufacturing but also the underlying supply chains. Yet, despite China's massive investments in a clean energy future, its policies have done more to promote its domestic economic interests and immediate environmental goals (such as reducing urban pollution) rather than building out supply chains that advance transparency, fair labor, accountability, and, ultimately, sustainability in the broadest sense. Those must be starting points for clean energy policymakers too.

Promote Policies to Ensure Responsible Production and Sourcing of Minerals for a Clean Energy Future

As the United States ramps up clean energy manufacturing, it will require expanding imports of raw minerals and processed materials from abroad. Surprisingly, while environmentalists are often equated with being "locavores" who fret over food miles and emphasize local production, some environmentalists are becoming globalists when it comes to minerals supplies. Leading US environmental groups have argued that a reliable supply of minerals depends upon "maintain[ing] our alliances with nations that source these minerals." They explain that sourcing minerals from abroad allows for greater efficiency, lower costs, and more diversified supply chains.[20]

But calls for a global trade in materials must be paired with plans to ensure that materials are sourced, processed, and traded responsibly—issues about which environmentalists have had less to say.[21] As the South American salars, the Congo's artisanal mines, the Peruvian Andes, and the northern reaches of Siberia make clear, both the history and the present-day situations in each of these mining regions have been very different, yet, in each case, the stakes are high for the communities that are on the front lines of resource extraction and processing important to a clean energy transition. Historically, foreign mining companies have reaped profits supplying resources to a global marketplace at the expense of the environment, workers, and local communities. The concerns that Green New Deal activists have raised about "extractivism" in the context of fossil fuels are just as urgent in the case of the materials needed for a clean energy future.

Since the start of the twenty-first century, as global production of minerals boomed and concerns about the "resource curse" and extractivism began to mount, a thicket of nongovernmental and industry initiatives emerged all aimed at advancing sustainable development in the context of the mining industry. New programs adopted by international lending institutions, corporate social stewardship programs, and public-private regulatory partnerships aimed to change the landscape of mining. Many companies comply with the International Organization for Standardization standards for environmental management systems (ISO 14001). A growing number of companies report on their sustainability initiatives following protocols established by the Global Reporting Initiative. Visiting almost any mining company's website reveals extensive materials—testimonials, case studies, and reports—highlighting their commitment to sustainability and community engagement.

The rise of such regulatory initiatives reflects a broader transition in global governance away from state-led regulations toward voluntary industry-led self-reporting. There is a general agreement among scholars that transparency and reporting can play an important role in improving oversight of transnational corporations.[22] But, simultaneously, there are concerns that initiatives such as the Bettercoal Code, Towards Sustainable Mining, or the International Council on Mining and Metals' Sustainable Development Framework—all industry-led initiatives—serve more to legitimize the practices of extractive industry than to reform them. Despite the substantive principles and provisions for reporting, these frameworks emphasize corporate self-regulation and generally include weak provisions for accountability and assessment.

Improving the sustainability of the mining and materials sectors both in practice and in ways that are broadly recognized is going to require a more collaborative approach. The Initiative for Responsible Mining Assurance offers one such strategy. Over the past decade, it has convened a participatory process engaging industry, organized labor, nongovernmental organizations, and mining communities to jointly establish the Standard for Responsible Mining, which is finalized, and a Chain of Custody Standard, which is in the works.

The Standard for Responsible Mining is a certification system that assesses mining activities relative to best practices in categories including worker health and safety, human rights, corruption, community engagement, pollution control, and land reclamation through mandatory third-party audits and publicly available scorecards.[23] Development of the standards was an iterative process that engaged a wide range of stakeholders and encouraged public input. Instead of simply showcasing corporate sustainability initiatives, as do many existing initiatives, the Standard for Responsible Mining requires a third-party, quantitative assessment of sustainability performance at mine sites across a wide range of metrics. That approach provides a basis for comparison, assessing change over time, and prioritizing areas for improvement. The Chain of Custody Standard, which is being finalized, aims to make it possible for downstream users of minerals and materials to track the provenance of materials in their possession and products.[24]

If there are going to be systems in place to assess and track the provenance of the materials that enable a clean energy future, they are likely to look a lot like the independently established and assessed standards that the Initiative for Responsible Mining Assurance has developed. But just as the scale of materials production is set to grow rapidly, that growth will need to be matched by the adoption and implementation of such standards. As of 2021, the standards remained largely in the piloting stage. Responsibility for complying with these standards cannot rest solely with mining companies. If such standards are going to make a difference, it will require that others—governments, financial lenders, downstream buyers, environmental groups, and consumers—support the adoption and implementation of such standards by mining companies and demand the integration of such standards into policies that govern domestic mining operations and international trade. Such transparency is essential, yet it is almost entirely nonexistent in the supply chains important to the battery industry—whether for lead-acid batteries, disposable batteries, or lithium-ion batteries—or other technologies important to a clean energy future.

*Support New and Expanded Mining and Refining Operations
in the United States*

In recent years, champions of the United States mining industry have urged the nation to roll back regulations and ramp up domestic mining to strengthen American "minerals independence," in part to support the clean energy transition.[25] The centerpiece of this strategy has been the American Mineral Security Act, which aims to advance that agenda by streamlining environmental regulations and expediting permitting for domestic mining development.[26] The legislation is strongly opposed by some of the same environmental groups that strongly support a clean energy transition and emphasize the importance of a global supply of minerals.[27] But, just as the United States has a responsibility to reduce its greenhouse gas emissions to slow climate change, it has a responsibility to materially support a global transition to a clean energy future too.

This will mean a turnabout in the development and processing of mineral resources in the United States. Since the mid-twentieth century, the United States has gone from being a leader in global minerals production to being a leader in global minerals consumption. While consumption of key metals—copper, aluminum, lead, and zinc—has been relatively flat since 1990, primary domestic production of those metals has fallen by 70 percent. That is a continuation of a trend that started in the 1940s, as domestic mining fell and import reliance increased. Even when factoring in recycling, production has still fallen by 40 percent.[28] This shift is a product of many intersecting factors, including the demand for a greater variety of minerals (more of which are mined outside the United States), tightening domestic regulations, international development policies, and global trade.

The lithium industry reflects this shift. The United States led the world in lithium production through the early 1990s. But as demand for lithium-ion batteries began to boom, low-cost lithium production surged first in South America and then Australia. By 2019, the United States accounted for less than 1 percent of global minerals production for lithium-ion batteries and roughly 1 percent of materials refining and processing; in contrast, China accounts for roughly 20 percent of production and 80 percent of processing globally.[29] Although the loudest voices championing the rejuvenation of the American mining industry hold up prospects for "minerals independence," the United States is very unlikely to be a major producer of many minerals critical to batteries, including nickel, manganese, and cobalt. But the United States could make nontrivial contributions to other materials, such as copper, lithium, and graphite.

Environmentalists raise the alarm about expanding domestic mining operations for good reason. Mining is a toxic industry that threatens what the Green New Deal rightly emphasizes are "front-line and vulnerable communities," which have borne a disproportionate burden of pollution and ill health historically. Environmentalists are right to press for much-needed regulatory reforms to ensure that mining operations are sited in consultation with local communities, operated safely, pay fair-market royalties when developed on public or tribal lands, and are prepared to undertake future remediation. Indeed, long-proposed domestic legislation such as the Hardrock Leasing and Reclamation Act would advance such policies.[30] But those proposed reforms need to be paired with clear goals to expand mining and materials processing in the United States in support of a transition to a clean energy future.

This will require a shift in strategy. Only a few environmental groups have begun to size up the material implications of the clean energy transition. At the forefront in the United States is Earthworks, a small environmental group with a long history of watchdogging fossil fuel and mineral extraction. In 2019, it launched a campaign titled "Making Clean Energy Clean, Just and Equitable." As it acknowledges, there is a need to "dramatically accelerate the transition to clean, renewable energy sources while simultaneously ensuring that people and the environment are not put at risk through destructive mining."[31] It emphasized the need to boost recycling, ensure responsible mineral sourcing, and shift consumption to lower demand. In 2020, the Sierra Club adopted a mining policy, including minerals for clean energy technologies. It described the challenge of sourcing minerals as a "global environmental and human rights challenge that needs to be resolved globally." In that context, it emphasized that the United States cannot rely on the rest of the world for minerals while restricting their development at home.[32]

Supporting the expansion of mining domestically will not mean greenlighting all proposed mines, but it will mean supporting more mining, expediting the review of mining proposals, and ensuring that mines are developed in ways that create economic opportunities for and advance the interests of local communities. Such policies only make sense in the context of broader policy commitments that ensure a rapid transition to a clean energy future on the scale of a Green New Deal.

Stop Trying to Recycle Our Way to a Clean Energy Future

Since the early 2000s, there have been growing calls to address resource depletion, waste management, and recycling in the context of a "circular economy." In place of a linear economy, in which resources are used and discarded as they were in the twentieth century, the circular economy aims to close the loop on the material economy, eliminate waste, and lessen the impact of human activities on the environment. The basic ideas underpinning the "circular economy" are not new; many were articulated in *Cradle to Cradle*, a well-known manifesto for closing the loop through better industrial design, in 2002.[33] And, without question, a circular economy is an important goal. But, based on the history of and prospects for batteries, two concerns arise: one is that a circular economy often elevates recycling as an inherently clean and sustainable process and, second, focusing on recycling distracts attention from the scale of material demand needed to support a clean energy transition.

Let us begin with the lead-acid battery. In many respects, lead-acid batteries appear to offer a model for a circular economy. Lead-acid batteries are the most highly recycled product in the world. In developed countries, almost all old lead-acid batteries are recycled and nearly all of the materials in spent batteries are recaptured for reuse in new batteries. But lead-acid batteries also offer a sharp reminder of the risks inherent in recycling. Both in developed countries and developing countries, ill-managed lead-acid battery recycling operations have been grave sources of harm for public health and the environment. At large-scale secondary lead smelters, like those in the United States and, far more problematic, smaller-scale facilities in poorer countries, recycling lead-acid batteries remains a risky enterprise if not managed carefully. Recent research estimates that one in three children globally have blood lead levels that threaten their development; informal lead-acid battery recycling in developing countries is a leading contributor to this epidemic.[34] In developing countries, immediate investments to modernize lead recycling facilities are imperative. More broadly, we need the same kinds of carefully researched and collaboratively designed standards to certify recycled materials as the Initiative for Responsible Mining Assurance has developed for sourcing newly mined materials. Already, the Initiative has identified this as an area for future standards setting.

In some cases, however, the costs of recycling may outweigh the benefits. On this point, the history of AA, AAA, and other single-use batteries is illustrative. Historically, when single-use batteries have been recycled, the recovered materials have usually been down-cycled into fertilizers, cement aggregate, and other lower-quality end products. Despite long-standing efforts to recycle single-use batteries, by most measures the costs of recycling are generally greater than the benefits of doing so (in part, because it takes so much energy to collect and transport spent batteries and the materials used to make new disposable batteries are generally abundant).[35] Although companies such as Energizer have achieved higher-quality recycling, turning old single-use batteries into feedstock for new batteries, the environmental costs of doing so remain unclear, making it difficult to determine how beneficial those processes are. The lesson is that instead of assuming that recycled materials are intrinsically more sustainable, those claims must be demonstrated. This is important in the case of lithium-ion batteries too. Improvements in the scale, process, and economics will be necessary for lithium-ion battery recycling to approach the efficiency of the lead-acid battery industry.

Even with well-regulated and highly efficient recycling processes, however, an emphasis on building a circular economy cannot distract attention from the challenge of scaling up minerals and materials production needed to build a clean energy economy. This is a risk. In 2019, the World Bank launched a "Climate-Smart Mining" initiative to support "the sustainable extraction and processing of minerals and metals used in clean energy technologies."[36] It aimed to help resource-rich developing countries expand their economies, to invest in mining technologies that reduced environmental and climate impacts, and to reduce the consequences of mining activities for "vulnerable communities." The World Bank's checkered history with respect to sustainable development and poverty reduction globally meant the initiative drew sharp scrutiny. The response of many environmentalists was telling. In a letter to the World Bank, while environmental leaders acknowledged the need for minerals and metals to deploy clean energy technologies, their primary concern was that the World Bank's program promoted new mining investments first, instead of making investments in recycling, promoting strategies to reduce demand (in the case of electric cars, this would mean promoting electrified public transit instead of private vehicles), or making what they described as "other non-mining solutions" the key components of a "Climate Smart" agenda.[37]

While those are all important goals, they fail to grapple with the immediate reality. To contend that the world can rely on mining operations only when "absolutely necessary"—as the environmentalists put it—is problematic. For the world to have a chance at weaning itself off fossil fuels quickly enough to meet the climate challenge, it is apparent that no amount of recycling can meet the material demands of a clean energy future—there is not nearly enough nickel, graphite, lithium, or other materials in circulation and ready to be recycled. And it is highly unlikely that policies to promote public transit, energy conservation, and other shifts in consumption can reduce demand at a scale nearly large enough to offset demand for new metals and minerals. Projections are that, even if the world fully closes the loop on lithium-ion batteries, recovered materials will reduce the total need for raw materials for lithium-ion batteries by 20 to 30 percent through 2050.[38] In short, although a circular economy is an essential long-term goal, the near-term challenge is ensuring that the supply of raw materials needed for batteries and other clean energy technologies is expanded significantly, which will mean substantial investments in mining and resource extraction, in ways that are both responsible and sustainable.

I realize there is an unasked question here, and my students would be the first to point it out. Isn't this all a pipe dream? Will a transition to a clean energy future, even if bolstered by these priorities, really matter if we don't also rethink consumerism, curb consumption, and rein in global capitalism? To put this in more immediate terms, isn't buying a battery-powered electric car really just buying into the myths of green consumerism, which have made it all too easy for privileged, well-meaning environmentalists (like myself) to ignore the root problems with modern consumerism and the global economy on which it turns? I've put up solar panels, replaced my oil furnace with electric heat pumps, and retired our aging Prius for an all-electric car.

That makes me one of those environmentalists who feels like I'm helping to save the planet when, in fact, my investments in a clean energy future are just perpetuating environmental havoc. Maybe, as the writer Jenny Price argues, it is time to stop trying to save the planet. As she puts it, we need to own up to the reality that a Prius or, more recently, a Tesla, "contributes a lot more than it doesn't to the wildly high-polluting industrial practices that devastate our atmosphere, bodies, forests, rivers, oceans, wetlands, wildlife, and more." As she reminds her eco-minded readers, those Priuses and Teslas aren't flying into the sky at night and magically scrubbing carbon out of the atmosphere.

No, they aren't. But we've only got years, not decades, to turn this planet-size ship around. That isn't much time to reinvent the profit-maximizing systems of capital accumulation and modern consumerism at the same time that we pull billions of people out of energy poverty and transition the world to renewable energy. Sure, we can fret about the 1 percent who are buying Teslas today, but keep in mind that in the early 1990s, cell phones represented the height of conspicuous consumption. Now, half the people in the world own a smartphone.[39] I, for one, am casting my lot with those who think our best bet is to reform these systems in ways that measurably advance both sustainability and social justice, while expanding access to energy and mobility.

In making this bet, I am not forsaking my deep concerns for the vitality, beauty, and wonder of the planet or a sense of urgency around addressing the deep inequities that stem from modern consumerism and global capitalism. But, as we prepare for a rapid transition to a clean energy future on the scale and with the promise of that envisioned by the Green New Deal, we must be aware that this future could still unfold in vastly different ways. It is very possible that such a transition, despite its aspirations for sustainability and equity, could reproduce the injustices of the fossil fuel–powered twentieth century: ignoring issues of equity, concentrating power in the hands of the corporate elite, and furthering extractivist policies that jeopardize both environments and frontline communities.

Some thinkers and policymakers have already begun mapping out a path to ensure that the energy transition is as much about energy democracy as it is about renewable energy.[40] It is possible to imagine a future in which inexpensive electric cars and buses and cheaper forms of micromobility are readily available to everyone.[41] And while solar panels are still a significant investment, the plummeting price of renewables is already making it possible for less-developed countries to build smaller-scale, more resilient, and more democratic systems of power distribution that leapfrog the centralized, fossil fuel–powered utilities that are hamstringing utility reform and climate action in countries like the United States.[42]

Make no mistake: this is not to suggest that the route to a more sustainable future centers on eco-minded individuals simply purchasing electric cars and solar panels. It means adopting policies that rapidly scale up those technologies, making them inexpensive and readily available globally. And, just as important, it means that those of us most concerned with a just and sustainable future need to lean into a more material environmentalism, shorn of the antimodern ideals that have limited environmental thought. It means not just supporting policies that promote energy efficiency and renewable energy but also supporting policies to ensure the underlying technologies are sourced and manufactured safely. Only then can we begin to address the consequences of those entanglements for workers, communities, and landscapes globally. In short, to return to the immediate climate challenge, the imperative is not just transitioning away from fossil fuels but, rather, transitioning to a clean energy future that addresses issues of social justice and sustainability.

And that brings me back to my core argument: as urgent are the goals for reducing greenhouse gas emissions and addressing climate change, equally pressing is *how* the world will go about making this transition. As proponents of the Green New Deal have argued: "Transition is inevitable. Justice is not." But building a regenerative economy for the future is going to require more than just empowering local communities, relocalizing production, recycling resources, and creating systems that "work against and transform current and historic social inequities"—all key principles of a "just transition."[43] To scale up a clean energy future as quickly and at the scale necessary to avoid the worst consequences of climate change is going to require remaking the extractive economy to supply massive amounts of materials sourced, refined, and manufactured all around the globe, including in the United States.

This is both an enormous challenge and an enormous opportunity. The risks are clear: scaling up a clean energy future could very easily drive social inequities and concentrate economic power, just as the fossil fuel–powered extractive economy of the twentieth century did. But it is also an opportunity. For just as the infrastructure needed for a clean energy future has not yet been built, neither have most of the mines, factories, and other links in the supply chain that are needed to support that transition. That means solving the clean energy equation will be about far more than just achieving "net zero" carbon. It is going to require sustained attention to the material and economic bases of a clean energy transition. It is going to require global trade policies that harmonize standards for the production and trade in materials, goods, and waste that protect both workers and the environment. It is going to require new strategies that make it possible to trace the provenance of materials from mine to product. It is going to require that environmental groups champion not just clean energy goals but goals for scaling up resource extraction that meet high standards for sustainability. And, finally, it is going to require activists, consumers, and corporations demanding transparency and accountability along the supply chain.

All of this is crucial to building a just and sustainable clean energy future, from the ground up.

NOTES

INTRODUCTION

1 "State Electricity Profiles: Colorado," Energy Information Administration, accessed June 1, 2001, www.eia.gov/state/print.php?sid=CO.

2 "Clean Jobs Colorado, 2019," E2, September 2019, www.e2.org/wp-content/uploads/2019/09/E2-Clean-Jobs-Colorado-2019.pdf.

3 Energy Information Administration, *Annual Energy Outlook 1999 with Projections to 2020*, January 2000, 129.

4 Energy Information Administration, *Electricity Generation from Selected Fuels, AEO2021 Reference Case*, www.eia.gov/outlooks/aeo/electricity/excel/subtopic 3_fig1.xlsx.

5 Energy Information Administration, *Electricity Generation from Selected Fuels, AEO2021 Reference Case*.

6 Christopher Mims, "The Three Stumbling Blocks to a Solar-Powered Nation," *Wall Street Journal*, January 14, 2018.

7 Sean Fleming, "China Joins List of Nations Banning the Sale of Old-Style Fossil-Fueled Vehicles," *World Economic Forum*, November 11, 2020, www.weforum.org/agenda/2020/11/china-bans-fossil-fuel-vehicles-electric/.

8 Amory B. Lovins, *Soft Energy Paths: Toward a Durable Peace* (New York: Harper and Row, 1977).

9 Stephen Pacala and Robert Socolow, "Stabilization Wedges: Solving the Climate Problem for the Next 50 Years with Current Technologies," *Science* 305, no. 5686 (August 13, 2004): 968.

10 Mark Z. Jacobson and Mark A. Delucchi, "Providing All Global Energy with Wind, Water, and Solar Power, Part II: Reliability, System and Transmission Costs, and Policies," *Energy Policy* 39, no. 3 (March 2011): 1170–90.

11 "100% Colorado," The Solutions Project, accessed June 1, 2021, https://thesolutions project.org/wp-content/uploads/wce/state_Colorado.pdf.

12 Donald Worster, *The Wealth of Nature: Environmental History and the Ecological Imagination* (New York: Oxford University Press, 1993), 143.

13 "About Tesla | Tesla," www.tesla.com/about.

14 Alex Nikolai Steffen, "The Next Green Revolution," *Wired*, May 1, 2006.

15 Michael Shellenberger, Ted Nordhaus, et al., "Ecomodernist Manifesto: A Manifesto for a Good Anthropocene," 2015, www.ecomodernism.org.

16 David A. Kirsch, *The Electric Vehicle and the Burden of History* (New Brunswick, NJ: Rutgers University Press, 2000), 202.

17 Fred Schlachter, "No Moore's Law for Batteries," *Proceedings of the National Academy of Sciences* 110, no. 14 (April 2, 2013): 5273.

18 Matthew R. Shaner et al., "Geophysical Constraints on the Reliability of Solar and Wind Power in the United States," *Energy & Environmental Science* 11, no. 4 (April 18, 2018): 914–25.

19 Kirsten Hund et al., *The Mineral Intensity of the Clean Energy Transition* (Washington, DC: World Bank Group, 2020), 11.

20 Chengjian Xu et al., "Future Material Demand for Automotive Lithium-Based Batteries," *Communications Materials* 1, no. 1 (December 2020): 1–10.

21 Olivier Vidal, Bruno Goffé, and Nicholas Arndt, "Metals for a Low-Carbon Society," *Nature Geoscience* 6 (November 2013): 894–95.

22 John McNeill, *Something New under the Sun: An Environmental History of the Twentieth-Century World* (New York: Cambridge University Press, 2000), 15.

23 Cutler J. Cleveland, "Energy Transitions," in *The Encyclopedia of Earth*, September 1, 2013, www.eoearth.org/view/article/152561/. Christopher F. Jones explains the importance of energy quality in a historical context in *Routes of Power: Energy and Modern America* (Cambridge, MA: Harvard University Press, 2016), 198–200.

24 David E. Nye, *Consuming Power: A Social History of American Energies* (Cambridge, MA: MIT Press, 1998).

25 Jones, *Routes of Power*, 2.

26 A. W. Crosby, *Children of the Sun: A History of Humanity's Unappeasable Appetite for Energy* (New York: Norton, 2006), 162.

27 Richard F. Hirsh and Benjamin K. Sovacool, "Wind Turbines and Invisible Technology: Unarticulated Reasons for Local Opposition to Wind Energy," *Technology and Culture* 54, no. 4 (2013): 705–34.

28 Adam Rome, "Crude Reality," *Modern American History* 1, no. 1 (January 5, 2018): 1–6.

29 John Urry, *Mobilities* (Malden, MA: Polity, 2007), 272.

30 Michelle Mascarenhas-Swan, "The Case for a Just Transition," in *Energy Democracy: Advancing Equity in Clean Energy Solutions*, ed. Denise Fairchild (Washington, DC: Island, 2017).

31 At the forefront of such analysis is Shalanda Baker, *Revolutionary Power: An Activist's Guide to the Energy Transition* (Washington, DC: Island, 2021).

32 Kate Aronoff et al., *A Planet to Win: Why We Need a Green New Deal* (London: Verso, 2019), 156.

33 Joeri Rogelj et al., "Mitigation Pathways Compatible with 1.5°C in the Context of Sustainable Development," in *Global Warming of 1.5°C. An IPCC Special Report* (N.p.: United Nations Intergovernmental Panel on Climate Change, 2018), 95.

34 Some excellent books on batteries include R. H. Schallenberg, *Bottled Energy: Electrical Engineering and the Evolution of Chemical Energy Storage* (Philadelphia: American Philosophical Society, 1982); Seth Fletcher, *Bottled Lightning: Superbatteries, Electric Cars, and the New Lithium Economy* (New York: Hill and Wang, 2011); H. R. Schlesinger, *The Battery: How Portable Power Sparked a Technological Revolution* (Washington, DC: Smithsonian, 2010); Isidor Buchmann, *Batteries in a Portable World: A Handbook on Rechargeable Batteries for Non-Engineers* (Richmond, British Columbia: Cadex Electronics, 2011); and Steve Levine, *The Powerhouse: America, China, and the Great Battery War* (New York: Penguin, 2016).

35 Eric S. Hintz, "Portable Power: Inventor Samuel Ruben and the Birth of Duracell," *Technology and Culture* 50, no. 1 (January 2009): 24–57.

36 David Orr, *Ecological Literacy: Education and the Transition to a Postmodern World* (Albany: State University of New York Press, 1992), 4.

1. LEAD-ACID BATTERIES AND A CULTURE OF MOBILITY

1 Jeff Bartlett, "Your Electric Car Is Here: Is It As Good As Promised?," *Motor Trend*, June 1, 1996.

2 Oliver Staley, "The General Motors CEO Who Killed the Original Electric Car Is Now in the Electric Car Business," *Quartz*, April 7, 2017.

3 R. H. Schallenberg, *Bottled Energy: Electrical Engineering and the Evolution of Chemical Energy Storage* (Philadelphia: American Philosophical Society, 1982).

4 A. E. Watson, *Storage Batteries* (Lynn, MA: Bubier, 1908).

5 Francis Bacon Crocker et al., *Storage Batteries: A Practical Presentation of the Principles of Action, Construction, and Maintenance of Lead and Non-Lead Batteries and Their Principal Commercial Application* (Chicago: American Technical Society, 1935), 7.

6 Kirsch, *The Electric Vehicle and the Burden of History*.

7 Virginia Scharff, *Taking the Wheel: Women and the Coming of the Motor Age* (Albuquerque: University of New Mexico Press, 1992), 37.

8 Schallenberg, *Bottled Energy*, 286–87.

9 V. W. Pagé, *Storage Batteries Simplified, Operating Principles, and Industrial Applications* (New York: Norman W. Henley, 1917), 122.

10 H. Roberts, "Changing Patterns in Global Lead Supply and Demand," *Journal of Power Sources* 116, no. 1–2 (July 1, 2003): 23–31.

11 W.S.B. Company, *The Starting and Lighting Battery* (Cleveland, OH: Willard Storage Battery Company, 1928).

12 W.S.B. Company, *The Starting and Lighting Battery*, 24.

13 Edward R. Parker, "Sulphur and Pyrite," in *Mineral Resources of the United States, 1900* (Washington, DC: GPO, 1901), 817–18.

14 Booker T. Washington, "The Man Farthest Down: Child Labor and the Sulphur Mines," *Outlook*, June 17, 1911, 342–48.

15 "New Method of Mining Sulphur," *Assumption Pioneer* 39, no. 4 (July 11, 1896): 1.

16 Louis August Lynn, "The Sulphur Industry of Calcasieu Parish" (master's thesis, Louisiana State University, 1950).

17 "Mr. Curtis on Caddo Fields," *Shreveport Times*, March 20, 1909, 5.

18 Edward W. Parker, "Sulphur," in *Mineral Resources of the United States, 1910* (Washington, DC: GPO, 1911), 101; G. F. Loughlin, *Mineral Resources of the United States, 1920* (Washington, DC: GPO, 1921), 110a; Larry Foulk, "The Freeport Sulphur Company Enterprise at Grande Ecaille and the Rise and Fall of Port Sulphur, Louisiana," *Louisiana History: The Journal of the Louisiana Historical Association* 50, no. 2 (2009): 203–12; Lynn, "The Sulphur Industry of Calcasieu Parish," chap. 3.

19 "Antimony—Production, Markets, and Uses," *Raw Material*, 1920, 153.

20 George Vinal, *Storage Batteries: A General Treatise on the Physics and Chemistry of Secondary Batteries and Their Engineering Applications*, 3rd ed. (New York: Wiley and Sons, 1940).

21 Sandra Beebe, *The Company by the Bay: A Portrait of Edward S. Evans and the People of the Evans Products Company of Coos Bay, Oregon, 1928–1962* (Springfield, OR: Bay Press, 1988).

22 Schallenberg, *Bottled Energy*, 299–300.

23 M. F. Chubb and H. R. Harner, "The Effect of Temperature and Rate of Discharge on the Capacity of Lead-Acid Storage Batteries," *Transactions of the Electrochemical Society* 68, no. 1 (October 1935): 257.

24 W.S.B. Company, *The Starting and Lighting Battery*.

25 Christopher C. Sellers, *Hazards of the Job: From Industrial Disease to Environmental Health Science* (Chapel Hill: University of North Carolina Press, 1999).

26 Albert Russell et al., *Lead Poisoning in a Storage Battery Plant* (Washington, DC: GPO, 1933), 35.

27 Committee on Storage Batteries, "Rules for the Prevention of Lead Poisoning in the Maintenance of Storage Batteries," *Safety* 2, no. 9 (October 1914): 219–20.

28 Christian Warren, *Brush with Death: A Social History of Lead Poisoning* (Baltimore: Johns Hopkins University Press, 2000), 71.

29 Alice Hamilton, "Lead Poisoning in the Manufacture of Storage Batteries" (Washington, DC: GPO, 1914), 24.

30 Russell et al., *Lead Poisoning in a Storage Battery Plant*, table 3.

31 Sellers, *Hazards of the Job*; Warren, *Brush with Death*; Gerald Markowitz and David Rosner, *Deceit and Denial: The Deadly Politics of Industrial Pollution* (Berkeley: University of California Press, 2002).

32 Warren, *Brush with Death*, 129–32.

33 *Occupational Lead Exposure and Lead Poisoning* (New York: American Public Health Association, 1943), 21.

34 As quoted in Robert Faust, "Lead Belt Progressives: The Struggle for Social and Environmental Reform in Missouri Mining Communities" (Ph.D. diss., University of Missouri–Columbia, 2003), 64.

35 Charles Freeman Jackson, *Methods of Mining Disseminated Lead Ore at a Mine in the Southeast Missouri District* (Washington, DC: Bureau of Mines, 1929).

36 A. P. Watt, "Concentration Practice in Southeast Missouri," *Transactions of the American Institute of Mining Engineers* (1918): 348–51.

37 Timothy LeCain, *Mass Destruction: The Men and Giant Mines That Wired America and Scarred the Planet* (New Brunswick, NJ: Rutgers University Press, 2009), 164–68; R. S. Dean and P. M. Ambrose, *Development and Use of Certain Flotation Reagents* (Washington, DC: US Bureau of Mines, 1944).

38 David Mosby et al., *Final Phase I Damage Assessment Plan for Southeast Missouri Lead Mining District* (Missouri Department of Natural Resources, January 2009); Environmental Protection Agency, *Record of Decision: Big River Mine Tailings Superfund Site, St. Francois County, Missouri* (Kansas City, KS: Environmental Protection Agency, September 2011), 8.

39 Carl Wright as quoted in Faust, "Lead Belt Progressives," 42.

40 Assorted advertisements, *Waste Trade Journal* 28, no. 4 (November 22, 1924): 17–21.

41 American Bureau of Metal Statistics, *Lead Consumption in the United States*, 1930.

42 Carl A. Zimring, *Cash for Your Trash: Scrap Recycling in America* (New Brunswick, NJ: Rutgers University Press, 2005), 49.

43 Leif Fredrickson, "The Age of Lead: Metropolitan Change, Environmental Health, and Inner City Underdevelopment in Baltimore" (Ph.D. diss., University of Virginia, 2017), 59–63.

44 Fredrickson, "The Age of Lead," 60–70.

45 William P. Eckel, Michael B. Rabinowitz, and Gregory D. Foster, "Investigation of Unrecognized Former Secondary Lead Smelting Sites: Confirmation by Historical Sources and Elemental Ratios in Soil," *Environmental Pollution* 117, no. 2 (April 2002): 273–79. Eckel's follow-up research identified 660 sites.

46 Fredrickson, "The Age of Lead," 104.

47 Huntington Williams et al., "Lead Poisoning from the Burning of Battery Casings," *Journal of the American Medical Association* 100, no. 19 (May 1933): 1485–89; Warren, *Brush with Death*, 142–43.

48 For an excellent study of early lead poisoning, see Fredrickson, "The Age of Lead," chap. 2.

49 Williams et al., "Lead Poisoning from the Burning of Battery Casings."

50 Fredrickson, "The Age of Lead," chap. 2.

51 United States of America v. The Association of American Battery Manufacturers et al., Civil Action No. 6199, Western District of Missouri, U.S. District Court, February 6, 1950. Files related to this case are available through the Department of Justice.

52 Putnam, Hayes & Bartlett, *The Impacts of Lead Industry Economics on Battery Recycling* (Cambridge, MA, 1986), 27.

53 Si Wakesberg, ed., *Recycled Lead in the United States: A Study of Current Market Trends and Projections* (New York: National Association of Recycling Industries, 1975).

54 "A Test of Four Advertisements for 'Diehard' Batteries: Submitted to J. Walter Thompson Company, Chicago" (Croton-on-Hudson, NY: Ernest Dichter International Institute for Motivational Research, March 1968), box 94, Ernest Dichter Papers, Hagley Museum and Library, Wilmington, DE.

55 Willard Storage Battery Company, *Your Storage Battery: What It Is and How to Get the Most Out of It* (Cleveland, OH: Willard Storage Battery Co., 1915), 3.

56 David Gartman, *Auto-Opium: A Social History of American Automobile Design* (New York: Routledge, 2013), 155.

57 Bruce A. Jacobs, "Bob Kent Puts the Juice Back in Exide," *Industry Week*, September 3, 1979, 4.

58 David R. Prengaman, "Advanced Battery Grid Alloys," in *Proceedings of the Symposium on Advances in Lead-Acid Batteries: New Orleans, La., October 8–9, 1984*, ed. K. Bullock (Pennington, NJ: Electrochemical Society of America, 1984), 201–13.

59 "It Never Needs Water. Ever," advertisement, *Popular Mechanics*, October 1979, 165.

60 Art Koch, "New Grids, New Plates, New Battery Shapes," in *Battery Council International: 1981 Meeting* (Battery Council International, 1981), 29–31.

61 Robert Goodrich, "Getting the Lead Out: Missouri Exploration Expands," *St. Louis Post-Dispatch*, August 31, 1980.

62 Gerald Markowitz and David Rosner, *Lead Wars and the Fate of America's Children* (Berkeley: University of California Press, 2013), 52.

63 Lydia Denworth, *Toxic Truth: A Scientist, a Doctor, and the Battle over Lead* (Boston: Beacon, 2009); Lucas Reilly, "The Most Important Scientist You've Never Heard Of," Mental Floss, May 17, 2017, http://mentalfloss.com/article/94569/clair-patterson-scientist-who-determined-age-earth-and-then-saved-it.

64 Clair C. Patterson, "Contaminated and Natural Lead Environments of Man," *Archives of Environmental Health: An International Journal* 11, no. 3 (September 1, 1965): 344–60.

65 Robert A. Kehoe to Professor Harry V. Warren, April 12, 1965, Toxic Docs: Version 1.0 (New York: Columbia University and City University of New York, 2018), www.toxicdocs.org, MM95GRLrVwBE674J3L1EBe3by.

66 James E. Boudreau, "Memo Re: Response to Clair Patterson's Research," December 10, 1964, toxicdocs.org, NxyL517ZZnOZQ41Mnwxwk00g.

67 H. E. Hesselberg to Robert A. Kehoe, January 20, 1965, toxicdocs.org, YDEJqdR8zZmBEY46ZO8Lown5k.

68 Manfred Bowditch to Robert Kehoe, Lead Industries Association, December 26, 1957, toxicdocs.org, 2JGag1gw3vKeD7z2Rw005BpR6.

69 Philip J. Landrigan et al., "Epidemic Lead Absorption near an Ore Smelter," *New*

England Journal of Medicine 292, no. 3 (January 16, 1975): 123–29. I used data reported for children aged one to nine years old.

70 Philip J. Landrigan et al., "Neuropsychological Dysfunction in Children with Chronic Low-Level Lead Absorption," *Lancet* 305, no. 7909 (March 29, 1975): 708–12.

71 Herbert L. Needleman et al., "Deficits in Psychologic and Classroom Performance of Children with Elevated Dentine Lead Levels," *New England Journal of Medicine* 300, no. 13 (March 29, 1979): 689–95.

72 Markowitz and Rosner, *Lead Wars.*

73 *Occupational Safety and Health Act of 1969: Hearings Before the Select Subcommittee on Labor,* U.S. House of Representatives, 91st Cong., 1st Sess. (November 6, 1969) (statement of Walter J. Burke, Secretary-Treasurer of U.S. Steelworkers).

74 Harold J. Gibbons, Secretary-Treasurer, Teamsters Local 688, to Elliot Porter, Missouri Air Conservation Commission, May 18, 1967, toxicdocs.org, 93BD 785dJ7wOKYNeaJjb63nV.

75 Letter from the People of Herculaneum, December 19, 1969, toxicdocs.org, 93BD785dJ7wOKYNeaJjb63nV.

76 Patricia Herbert to US Senator Thomas F. Eagleton, June 9, 1970, toxicdocs.org, 93BD785dJ7wOKYNeaJjb63nV.

77 Chad Montrie, *The Myth of Silent Spring: Rethinking the Origins of American Environmentalism* (Berkeley: University of California Press, 2018).

78 As quoted in Henry Weinstein, "A Battery Plant and Lead Poisoning," *New York Times*, June 6, 1976. For more on the plant, see *Oversight Hearings on the Occupational Safety and Health Act: Hearings Before the Subcommittee on Manpower, Compensation, and Health and Safety of the Committee on Education and Labor,* U.S. House of Representatives, 94th Cong., 1st Sess. (1976) (statement of Hector P. Blejer, National Institute for Occupational Safety and Health).

79 Neil S. Shifrin, *Air Quality Management in the 20th Century and Doe Run's Herculaneum Lead Smelter Activities* (Cambridge, MA: Gradient Corporation, February 15, 2008).

80 Environmental Protection Agency, "Arcanum Iron & Metal, OH" (Washington, DC: EPA, September 26, 1986).

81 For a summary of regulations, see Putnam, Hayes, & Bartlett, *The Impacts of Lead Industry Economics and Hazardous Waste Regulation on Lead-Acid Battery Recycling* (1987), 16–17. In 1995 the EPA changed the RCRA policies under the Universal Waste Rule for widely generated hazardous wastes including lead-acid batteries destined for recycling.

82 Kenneth T. Wise et al., "The Economics of Lead-Acid Battery Recycling," *Proceedings of the Fourth National Conference on Waste Exchange* (March 1987): 97.

83 United States Geological Survey, *Historical Lead Statistics*, December 20, 2018, www.usgs.gov/centers/nmic/historical-statistics-mineral-and-material-commo dities-united-states.

84 Putnam, Hayes, & Bartlett, *The Impacts of Lead Industry Economics and Hazardous Waste Regulation on Lead-Acid Battery Recycling*, 22–26.

85 Norman Schroeder, "Small Secondary Lead Smelters Seen Cutting Environment Safeguard Cost," *American Metal Market*, November 15, 1988.

86 Putnam, Hayes, & Bartlett, *The Impacts of Lead Industry Economics*, 25.

87 James G. Palmer, GNB Incorporated, to Bob Wilbur, Battery Council International, July 10, 1986, in Putnam, Hayes, & Bartlett, *The Impacts of Lead Industry Economics*.

88 David Weinberg, "Washington Issues Update: Impact of EPA's Developing Lead Strategy," in *Battery Council International: Proceedings of 1990 Meeting* (San Francisco, 1990), 48–50.

89 "States' Efforts to Promote Lead-Acid Battery Recycling" (Washington, DC: Environmental Protection Agency, January 1992).

90 As quoted in Ted Kuster, "KMart, Exide Set Recycling Program," *American Metal Market*, May 24, 1990.

91 As quoted in Mikell Knights, "Environmental Rules Put a Cost Squeeze on the Lead Industry," *American Metal Market*, September 5, 1991.

92 J. W. Ropert, "Closed System," paper presented at the Battery Council International Convention (Mexico City, April 1976), 96–98; Michael Gardiner, "Secondary Lead Smelters and Clean Technology," in *Battery Council International: 1981 Meeting* (Battery Council International, 1981).

93 Andreas Siegmund, "Secondary Lead Smelting at the Beginning of the 21st Century," *Proceedings of the European Metallurgical Conference* (2001): 17.

94 J. L. Sullivan and L. Gaines, *A Review of Battery Life-Cycle Analysis: State of Knowledge and Critical Needs*, Argonne National Laboratory, October 1, 2010, 10.

95 Ropert, "Closed System."

96 USGS, *Historical Lead Statistics*, December 20, 2018.

97 Rick Godber, "BCI President's Perspective" in *Battery Council International: 1996 Meeting* (Battery Council International, 1996), 46–48.

98 Based on reports since 1990s from SmithBucklin Corporation, *National Recycling Rate Study* (Chicago) conducted for Battery Council International.

99 Jared E. Hazleton, *The Economics of the Sulphur Industry* (New York: Routledge, 1976).

100 Timothy Dignam et al., "Control of Lead Sources in the United States, 1970–2017: Public Health Progress and Current Challenges to Eliminating Lead Exposure," *Journal of Public Health Management and Practice* 25 (2019): S13–22.

101 Robert T. Pavlowky et al., *Distribution, Geochemistry, and Storage of Mining Sediment in Channel and Floodplain Deposits of the Big River System in St. Francois, Washington, and Jefferson Counties, Missouri* (Springfield: Ozarks Environmental and Water Resources Institute, Missouri State University, 2010).

102 Tom Uhlenbrock, "Lead Waste Cleanup Gets New Emphasis," *St. Louis Post-Dispatch*, December 26, 1993.

103 Agency for Toxic Substances and Disease Registry, *Big River Mine Tailings Superfund Site Lead Exposure Study, St. Francois County, Missouri* (Jefferson City: Missouri Department of Health, August 1998), 1.

104 Environmental Protection Agency, *Record of Decision: Big River Mine Tailings Superfund Site, St. Francois County, Missouri* (Kansas City, KS: EPA, Region VII, September 2011).

105 Patrick Blanks et al. vs. Fluor Corporation et al., No. ED97810 (Missouri Court of Appeals Eastern District, September 16, 2014), 45.

106 Shifrin, "Air Quality Management in the 20th Century and Doe Run's Herculaneum Lead Smelter Activities," 103–4.

107 Ken Midkiff and Sierra Club–Ozark Chapter to Chief, Air Planning Section, Missouri Department of Natural Resources, October 31, 2000, in Doe Run Company, June 2001 to September 2002, SPF Technical, Missouri Department of Natural Resources, Sunshine Files.

108 Missouri Department of Health and Senior Services, "Blood-Lead Level Health Consultation," February 26, 2002 (online document no longer available).

109 As quoted in Chris Carroll, "Tests Show Heavy Lead Poisoning in Herculaneum," *St. Louis Post-Dispatch*, February 28, 2002.

110 Washington University–St. Louis Interdisciplinary Environmental Clinic, "Fact Sheet: Proposed Revisions to the National Ambient Air Quality Standard for Lead," May 1, 2008, https://law.wustl.edu/intenv/documents/2008-05-01_factsheet.pdf.

111 Laura Gilcrest, "Lead Maker Doe Run to Close Missouri Smelter at End of 2013," *Platts Metals Week*, October 27, 2009.

112 William Eckel, "The Secondary Lead Smelting Industry" (Ph.D. diss., George Mason University, 2001).

113 Alison Young, "Long-Gone Lead Factories Leave Poisons in Nearby Yards," *USA Today*, April 19, 2012.

114 Edward Link, "Pre-CERCLIS Screening Assessment Checklist/Decision Form," September 26, 2003, 10, Ohio Environmental Protection Agency, www.documentcloud.org/documents/263120-site-131-tyroler-metals-ohio-epa-2003-and-2004.html#document/p12/a37994.

115 Ken Shefton as quoted in "Case Study: Tyroler Metals," *USA Today*, www.usatoday.com/videos/news/nation/2012/12/03/1743167/.

116 Office of Land and Emergency Management, "EPA Lead Smelter Strategy Summary Report" (Washington, DC: EPA, September 2017).

117 As quoted in Tony Barboza, "Exide's Hazardous Waste Permit Application Deficient, California Says," *Los Angeles Times*, June 17, 2014.

118 See photographs of protesters' signs included in Larry Buhl, "Burned by Slow Government Response to a Polluter, Residents Mistrust Cleanup Efforts," DeSmog, May 22, 2016.

119 "DTSC Announces Additional Requirements on Exide Plant," California Department of Toxic Substances Control, press release, October 7, 2013.

120 "Soil Sampling Results: Preliminary Investigation Area," California Department of Toxic Substances Control, October 14, 2016.

121 "Exide Technologies Admits Role in Major Hazardous Waste Case and Agrees to Permanently Close Battery Recycling Facility in Vernon," US Department of Justice, March 12, 2015; "Exide Closing Vernon Plant to Avoid Criminal Prosecution," *Los Angeles Times*, June 17, 2015.

122 On the global shift in the lead industry, and its hazards, see Christopher C. Sellers, "Cross-Nationalizing the History of Industrial Hazard," *Medical History* 54, no. 3 (2010): 315–40.

123 David E. Guberman, "Lead," in *2014 Minerals Yearbook* (Washington, DC: Bureau of Mines, November 2016).

124 Blacksmith Institute, *The World's Worst Polluted Places: The Top Ten* (New York, October 2006).

125 Empresa Minera Del Centro Del Peru S.A., *Preliminary Environmental Evaluation: Monitoring Report of Gaseous Effluents* (La Oroya, March 1995), 7, 10; Environmental Protection Agency, Toxic Release Inventory Data, Doe Run Herculaneum Smelter, 63048HRCLN881MA.

126 Ministry of Health, "Blood Lead Study on a Selected Group of the Population of La Oroya," Lima, Peru, November 1999. Note: Many of the documents related to La Oroya are drawn from *The Renco Group, Inc. v. Republic of Peru* (ICSID Case No. UNCT/13/1), https://icsid.worldbank.org/cases/case-database/case-detail?CaseNo=UNCT/13/1.

127 Corinne Schmidt, "How Brown Was My Valley," *Newsweek*, April 18, 1994, 21.

128 Ministry of Energy and Mines, Directorial Resolution No. 334-97-EM/DGM, October 16, 1997; Doe Run, Form S-4, Securities and Exchange Commission, May 11, 1998, www.sec.gov/Archives/edgar/data/1061112/0001047469-98-018990.txt.

129 Doe Run Peru, *Report to Our Communities in La Oroya* (2001), 8.

130 Doe Run Peru, *Report to our Communities in La Oroya* (2001), 77, 151.

131 Blacksmith Institute, *The World's Worst Polluted Places* (2006).

132 Sara Shipley Hiles and Marina Walker Guevara, "Lead Astray," *Mother Jones*, 2006, www.motherjones.com/politics/2006/10/lead-astray.

133 B. Hunter Farrell, "From Short-Term Mission to Global Discipleship: A Peruvian Case Study," *Missiology: An International Review* 41, no. 2 (April 1, 2013): 165.

134 Shayna Posses, "Renco Can't Shake Claims over Poisoning in Peru, Court Told," *Law360*, October 4, 2017.

135 Giovanna Gismondi, "The Renco Group, Inc. v. Republic of Peru: An Assessment of the Investor's Contentions in the Context of Environmental Degradation," *Harvard International Law Journal* 59 (2017): 12.

136 "Lead Poisoning Sickens 191 Children in Central China City," *New China News Agency*, March 24, 2010; Sharon Lafraniere, "Lead Poisoning in China: The Hidden Scourge," *New York Times*, June 15, 2011; Zhi Sun, "Spent Lead-Acid Battery Recycling in China—A Review and Sustainable Analyses on Mass Flow of Lead," *Waste Management* 64 (June 1, 2017): 190–201.

137 Bill D. Moyers, *Global Dumping Ground: The International Traffic in Hazardous Waste* (Washington, DC: Seven Locks, 1990).

138 Jonathan Krueger, *International Trade and the Basel Convention* (Washington, DC: Earthscan, 1999), xviii.

139 Michael Carr, Vice President and General Manager, North America Power Solutions, Johnson Controls, Inc. to Commission for Environmental Cooperation, June 26, 2012.

140 *Exporting Hazards: U.S. Shipments of Used Lead Batteries to Mexico Take Advantage of Lax Environmental and Worker Health Regulations* (San Francisco and Mexico City: Occupational Knowledge International and Fronteras Comunes, June 2011).

141 Mary Tiemann, *NAFTA: Related Environmental Issues and Initiatives* (Washington, DC: Congressional Research Service, 1999).

142 *Exporting Hazards: U.S. Shipments of Used Lead Batteries to Mexico Take Advantage of Lax Environmental and Worker Health Regulations.*

143 Elisabeth Rosenthal, "Recycled Battery Lead Puts Mexicans in Danger," *New York Times,* December 8, 2011.

144 Timothy Whitehouse, *Hazardous Trade? An Examination of US-Generated Spent Lead-Acid Battery Exports and Secondary Lead Recycling in Mexico, the United States and Canada* (Montreal, Quebec: Commission for Environmental Cooperation, April 2013).

145 Robert E. Finn and RSR Corporation to CEC re: Comments on Transboundary Movements of SLABs, June 8, 2012.

146 Shinsuke Tanaka, Kensuke Teshima, and Eric Verhoogen, "North-South Displacement Effects of Environmental Regulation: The Case of Battery Recycling," *American Economic Review: Insights* (2021), forthcoming.

147 Laura Coughlan, Eva Kreisler, and Jana Tatum, "The U.S. EPA Spent Lead Acid Battery Export Rule: Challenges of Implementing Regulations with Transnational Impact," June 2011 (online document no longer available).

148 "Dominican Republic—Bajos de Haina Abandoned Lead Smelter," *Pure Earth* blog, www.pureearth.org/project/haina/.

149 Pascal Haefliger et al., "Mass Lead Intoxication from Informal Used Lead-Acid Battery Recycling in Dakar, Senegal," *Environmental Health Perspectives* 117 (May 14, 2009): 1535–40.

150 "Phyllis Omido," Goldman Environmental Foundation, June 1, 2015, www.goldmanprize.org/recipient/phyllis-omido/.

151 M. Sim, "Fact Sheet: What Is the #1 Childhood Environmental Health Threat Globally?," Pure Earth, November 19, 2014.

152 Nicholas Rees and Richard Fuller, *The Toxic Truth: Children's Exposure to Lead Pollution Undermines a Generation of Future Potential* (UNICEF and Pure Earth, 2020); Alison L. Clunes et al., "Mapping Global Environmental Lead Poisoning in Children" *Blacksmith Institute Journal of Health and Pollution* 1, no. 2 (November 2011): 20.

153 Rapier, "Lead-Acid Batteries Are on a Path to Extinction," *Forbes,* October 13, 2019.

154 USGS, *Historical Lead Statistics,* December 20, 2018.

155 "Leading the Way in Battery Recycling," Johnson Controls, Inc., infographic, Milwaukee, WI, 2012.

156 Data from Toxic Release Inventory, Environmental Protection Agency, www .epa.gov/toxics-release-inventory-tri-program.

157 Pietro P. Lopes and Vojislav R. Stamenkovic, "Past, Present, and Future of Lead-Acid Batteries," *Science* 369, no. 6506 (August 21, 2020): 923–24.

158 Po-Yen Chen et al., "Environmentally Responsible Fabrication of Efficient Perovskite Solar Cells from Recycled Car Batteries," *Energy & Environmental Science* 7, no. 11 (October 15, 2014): 3659–65.

2. AA BATTERIES AND A THROWAWAY CULTURE

1 Elsa Olivetti, Jeremy Gregory, and Randolph Kirchain, *Life Cycle Impacts of Alkaline Batteries with a Focus on End-of-Life* (study for National Electrical Manufacturers Association, MIT, February 2011).

2 My calculations are based on data and analyses included in Olivetti, Gregory, and Kirchain, *Life Cycle Impacts of Alkaline Batteries with a Focus on End-of-Life*. Note that the calculations presented in the text should be treated as estimates, considering that the data underlying the Olivetti, Gregory, and Kirchain study was collected prior to 2011.

3 I assume that the capacity of an AA battery is approximately 4 watt-hours, the price of a battery is sixty-six cents, and the average price of electricity in the United States is thirteen cents per kilowatt-hour.

4 "Electrical Characteristics and Testing of Dry Cells," *General Electric Review* 22 (December 1919): 1011.

5 Fred de Land, "Notes on the Development of Telephone Service VI," *Popular Science Monthly* 70 (May 1907).

6 A. E. Dobbs, "Exchange Inspection," *Telephone Magazine*, March 1900.

7 Philip Nungesser to Gentlemen, August 1, 1914, box 5, folder "Nungesser Carbon & Battery," Soo Hardware Company Archives, Hagley Museum and Library, Wilmington, DE.

8 Nungesser Price Sheet, October 1, 1913, box 5, folder "Nungesser Carbon & Battery," Soo Hardware Company Archives, Hagley Museum and Library, Wilmington, DE; "Nungesser Advertisement, *Telephony*, November 13, 1915, 11.

9 "Electrical Characteristics and Testing of Dry Cells," *General Electric Review* (1919): 1011.

10 Trent E. Boggess and Ronald Patterson, "The Model T Ignition Coil," *Vintage Ford Magazine* 34 (1999): 21.

11 "French Ray-O-Sparks Make Better Motors," advertisement for farm publication, 1924.

12 James G. Harbord, "Radio," *North American Review* 231, no. 6 (1931): 531–34.

13 Steve Craig, "'The More They Listen, the More They Buy': Radio and the Modernizing of Rural America, 1930–1939," *Agricultural History* 80, no. 1 (2006): 1–16.

14 Austin C. Lescarboura, *Radio for Everybody* (New York: Scientific American, 1925).

15 "Small Current—Big Job!," *Wireless Age*, February 1923, 7.

16 *101 Uses for an Eveready* (New York: American Ever Ready Works, 1917), 3.

17 Eugene H. Mathews, "Flashlights and Flashlight Batteries," in *Illuminating Engineering* (Baltimore: Illuminating Engineering Society, 1922), 137.

18 *101 Uses for an Eveready.*

19 Mathews, "Flashlights and Flashlight Batteries."

20 "Standardization of Dry Cells," *Electrochemical and Metallurgical Industry* 7, no. 12 (December 1909): 501.

21 Charles F. Burgess, "Desirable Characteristics of a Dry Cells," *Electrochemical and Metallurgical Industry* 7, no. 12 (December 1909): 524–25.

22 F. H. Lovebridge, "Dry Cells," *Electrochemical and Metallurgical Industry* 7, no. 12 (December 1909): 523–24.

23 US Bureau of Standards, *Electrical Characteristics and Testing of Dry Cells* (Washington, DC: Department of Commerce, 1919), table IV.

24 US Department of Commerce, *United States Government Master Specification for Cells and Batteries, Dry* (Washington, DC: US Department of Commerce, 1927).

25 US Bureau of Standards, *American Standard Specification for Dry Cells and Batteries* (Washington, DC: US Department of Commerce, 1937).

26 US Bureau of Standards, *American Standard Specification for Dry Cells and Batteries*, iv.

27 Ray-O-Vac, "Full Line Catalog," n.d., Rayovac Records.

28 Olivetti, Gregory, and Kirchain, *Life Cycle Impacts of Alkaline Batteries with a Focus on End-of-Life*, 2011.

29 Thomas S. Jones, "Manganese," in *Minerals Yearbook* (Washington, DC: Bureau of Mines, 1990).

30 For general background on manganese, see John Emsley, "Manganese the Protector," *Nature Chemistry* 5, no. 11 (November 2013): 978; Fraser A. Armstrong, "Why Did Nature Choose Manganese to Make Oxygen?," *Philosophical Transactions of the Royal Society B: Biological Sciences* 363, no. 1494 (March 27, 2008): 1263–70.

31 George W. Heise and N. C. Cahoon, "Fiftieth Anniversary: The Anniversary Issue on Primary Cell Systems: Dry Cells of the Leclanché Type, 1902–1952—A Review," *Journal of the Electrochemical Society* 99, no. 8 (August 1, 1952): 184C.

32 "Methods and Cost of Mining at Chiaturi," *Mining Journal, Railway, and Commercial Gazette*, July 30, 1898.

33 John B. Myer, President, Philipsburg Mining Company, to Board, March 18, 1916, box 1, folder 3, Philipsburg Mining Company Records [hereafter PMC Records], Montana Historical Society, Helena.

34 John B. Myer, President, Philipsburg Mining Company, to John R. Lucas, Superintendent, April 5, 1916, box 1, folder 3, PMC Records.

35 C. W. Storey, "Abstract of Report N-665 on Philipsburg, Montana, Ore," February 14, 1917, box 3, folder 18, PMC Records.

36 John B. Myer, President, Philipsburg Mining Company, to Board, August 18, 1917, box 1, folder 12, PMC Records.

37 William R. Cymer, Depolarizing Mixture and Method of Preparing Same, US Patent 1480533A, filed September 17, 1920, and issued January 8, 1924.

38 Heise and Cahoon, "Fiftieth Anniversary: The Anniversary Issue on Primary Cell Systems: Dry Cells of the Leclanché Type, 1902–1952—A Review," 179C.

39 United States Geological Survey, "Manganese," in *Minerals Yearbook* (Washington, DC: Bureau of Mines, 1924).

40 Superintendent to John B. Myer, President, Philipsburg Mining Company, March 14, 1921, box 1, folder 13, PMC Records.

41 George Vinal, *Primary Batteries* (New York: Wiley, 1950), 59.

42 Production statistics drawn from manganese coverage in *Minerals Yearbook* between 1920 and 1940.

43 *History of Ray-O-Vac Corporation*, 1965, Rayovac Records.

44 Ray-O-Vac, *Ray-O-Lite Sales Idea Book* (1924), box 2, Rayovac Records.

45 Ray-O-Vac, "New and Improved French Telephone Cells" (ca. 1920s), Rayovac Records.

46 US Bureau of Standards, *American Standard Specification for Dry Cells and Batteries.*

47 Douglas B. Grant, "Batteries Wanted in Iowa," *Broadcasting*, March 29, 1943.

48 George Raynor Thompson and Dixie Harris, *The Signal Corps: The Outcome* (Washington, DC: Center of Military History, US Army, 1966), 385.

49 US War Department, *Technical Manual: 2.36-inch A.T. Rocket Launcher* (1943), section II.

50 As quoted in Thompson and Harris, *The Signal Corps*, 222.

51 War Production Board, *War Production in 1944* (Washington, DC: War Production Board, June 1945), 60.

52 "The Wizard of New Rochelle," *Business Week*, November 4, 1967, 70.

53 Maurice Friedman and Charles E. McCauley, "The Ruben Cell—A New Alkaline Primary Dry Cell Battery: Its Design and Manufacture, Electrochemical Principles, Performance Characteristics, and Applications," *Transactions of the Electrochemical Society* 92, no. 1 (October 1, 1947): 195–215.

54 E. S. Hintz, "Portable Power: Inventor Samuel Ruben and the Birth of Duracell," *Technology and Culture* 50, no. 1 (January 2009): 33–35.

55 *Strategic and Critical Materials and Metals: Hearings Before the Subcommittee on Mines and Mining,* U.S. House of Representatives, 80th Cong., 2nd Sess. (May 20, 1948) (statement of S. H. Williston, Cordero Mining Co.), 912–13.

56 Helena A. Meyer and Alethea W. Mitchell, "Mercury" in *Minerals Yearbook, 1944* (Washington, DC: Bureau of Mines, 1946), 697.

57 "Mercury Gives a Miracle Battery," *Berkshire Eagle*, January 26, 1945.

58 Meyer and Mitchell, "Mercury," 717.

59 P. A. Sanitarian to Medical Director, memorandum re: report of the environmental investigation of P. R. Mallory Company, March 30, 1945, Rayovac Records.

60 "Tiny, New Mallory Battery Does Big Job in Tropics," *Indianapolis Star*, January 24, 1945; "Mercury Gives a Miracle Battery."

61 Hintz, "Portable Power."

62 Julie Michelle Klinger, *Rare Earth Frontiers: From Terrestrial Subsoils to Lunar Landscapes* (Ithaca, NY: Cornell University Press, 2018).

63 Megan Black, *The Global Interior: Mineral Frontiers and American Power* (Cambridge, MA: Harvard University Press, 2018), 123–29.

64 *Mineral Resources Development: Hearings Before the Committee on Interior and Insular Affairs*, U.S. Senate, 81st Cong., 1st Sess. (July 1949) (statement of James Boyd, Director, Bureau of Mines), 26.

65 Black, *The Global Interior*, 123.

66 United States President's Materials Policy Commission, *Resources for Freedom: A Report to the President* (Washington, DC: GPO, 1952), 1.

67 Black, *The Global Interior*, 128.

68 Church to Just, memorandum, Economic Cooperation Administration (October 28, 1948), box 155, Records of the US Foreign Assistance Agencies, 1948–61, Record Group 469, US National Archives, College Park, MD.

69 Harry F. Acheson, Memo Regarding Gold Coast Manganese (April 23, 1949), box 155, Records of the US Foreign Assistance Agencies, 1948–61, Record Group 469, US National Archives, College Park, MD.

70 Kenneth Douglas Ruble, *The RAYOVAC Story: The First 75 Years* (Madison, WI: Rayovac, 1981), 115.

71 *Strategic and Critical Materials and Metals Hearings* (March 10, 1948) (statement of J. Carson Adkerson, President of the American Manganese Producers Association), 202.

72 Ruble, *The RAYOVAC Story*, 115.

73 Gilbert L. DeHuff and Teresa Fratta, "Manganese," in *Minerals Yearbook* (Washington, DC: Bureau of Mines, 1955), 753.

74 Joseph C. Schumacher, "Ventures in Electrochemical Industry," *Journal of the Electrochemical Society* 129, no. 10 (October 1, 1982): 400C–401C.

75 "Effect of Electrolytic Manganese Dioxide (EMD) on Cell Performance," in *Batteries and Energy Systems*, ed. Charles Mantell (New York: McGraw-Hill, 1983).

76 J. Prabhakar Rethinaraj, "Preparation and Properties of Electrolytic Manganese-Dioxide," *Hydrometallurgy* 42, no. 3 (1993): 335–43; Olivetti, Gregory, and Kirchain, *Life Cycle Impacts of Alkaline Batteries with a Focus on End-of-Life*.

77 Kazuhide Miyazaki, "Historical Key Factors for Manufacture of Electrolytic Manganese Dioxide in Japan," in *Proceedings of the Symposium on History of Battery Technology*, ed. A. J. Salkand (Pennington, NJ: Electrochemical Society, 1987), 83–95.

78 Howard G. McEntee, "Clever New Gadgets Run on Flashlight Batteries," *Popular Science*, March 1959.

79 Chet Stephens, "How to Choose the Right Battery," *Electronics Illustrated*, November 1963, 103.

80 Ray-O-Vac, "Current Industrial Report—Dry Cell Batteries and Cases" (1961), box 1, Rayovac Records.

81 Alan Farnham, "What's Sparking Duracell?" *Fortune*, July 16, 1990.

82 Ruble, *The RAYOVAC Story*, 115.

83 Terry Telzrow, interview by author, June 21, 2013.

84 Robert Scarr, interview by author, July 22, 2013.

85 Mallory Corporation, "Advertisement: Duracell. Still Going Strong . . . ," *Popular Mechanics*, March 1970, 25.

86 R. J. Brodd, A. Kozawa, and K. V. Kordesch, "Primary Batteries 1951–1976," *Journal of the Electrochemical Society* 125, no. 7 (July 1, 1978): 272C.

87 "Sony Celebrates Walkman 20th Anniversary," Sony Global, July 1, 1999, www .sony.net/SonyInfo/News/Press_Archive/199907/99-059/.

88 Bob Greene, "Walkman Earphones: Mind-Altering Devices," *Salisbury (MD) Daily Times*, October 31, 1981.

89 Rebecca Tuhus-Dubrow, *Personal Stereo* (New York: Bloomsbury Academic, 2017); Rebecca Tuhus-Dubrow, "The Gadget That Broke Humanity," *Boston Globe*, September 1, 2017.

90 Jacqueline Simmons, "Rayovac Sets $20 Million Push to Power Rechargeable Battery," *Wall Street Journal*, December 23, 1993.

91 Theodore Steinberg, *Down to Earth: Nature's Role in American History* (New York: Oxford University Press, 2002), chap. 14.

92 *Resource Conservation and Recovery Act of 1976: Hearings Before the Subcommittee on Transportation and Commerce of the Committee on Interstate and Foreign Commerce,* U.S. House of Representatives, 94th Cong., 2nd Sess. (June 1976) (statement of Fred B. Rooney), 1.

93 Steinberg, *Down to Earth*, 233–34.

94 Andrew Pollack, "Battery Pollution Worries Japanese," *New York Times*, June 25, 1984.

95 Deidre Lord, "Burnt Out Batteries," *Environmental Action*, September/October 1988.

96 David L. Wojichowski, *Source of Heavy Metals in Municipal Solid Waste: Mercury, Cadmium, and Lead* (Des Plaines, IL: Signal Environmental Systems, October 7, 1986), 3.

97 *Background Information on Hennepin County's Plan for Solid Waste* (Minneapolis, MN: Board of Hennepin County Commissioners, 1988), 2.

98 Allen Hershkowitz, "Burning Trash: How It Could Work," *Technology Review*, July 26, 1987.

99 Patrick L. Reagan, Howard W. Mielke, and David Stoppel, "The City at Risk: Lead (Pb) Emissions from the Minneapolis Garbage Smelter" (St. Paul, MN: Lead Coalition, May 17, 1988).

100 Edward B. Swain, "Assessment of Mercury Contamination in Selected Minnesota Lakes and Streams," box 3, Minnesota Pollution Control Agency Published Records, Minnesota State Archives, St. Paul.

101 Kevin O'Donnell, "Memo Re: Meeting with Senator Dahl on Household Batteries" (October 24, 1988), box 9, Waste Management Board, Hazardous Waste Facilities Files, Minnesota State Archives; "Notes on Battery Legislation

Meeting" (December 6, 1989), box 4, Legislative Commission on Waste Management, Administrative Records, Minnesota State Archives.

102 Kevin Taylor, David J. Hurd, and Brian Rohan, "Recycling in the 1980s: Batteries Not Included," *Resource Recycling*, May/June 1988, 26–27.

103 Lord, "Burnt Out Batteries," 18.

104 Raymond L. Balfour, "Household Battery Disposal: An Overview," in *First International Seminar on Battery Waste Management: A Three-Day Seminar and Workshop*, ed. S. Wolsky (1989).

105 Minnesota Statutes §115A.9155 and §325E.125 (April 5, 1990).

106 As quoted in Steve Brandt, "Battery Law Sets Rules for Makers, Users," *Minneapolis Star Tribune*, April 5, 1990.

107 Jennifer Nash and Christopher Bosso, "Extended Producer Responsibility in the United States: Full Speed Ahead?," *Journal of Industrial Ecology* 17, no. 2 (April 2013): 175–85.

108 *Implementation of the Mercury-Containing and Rechargeable Battery Management Act* (Washington, DC: EPA, November 1997).

109 Jean-Yves Huot, interview by author, July 31, 2013.

110 National Electrical Manufacturers Association, "Mercury Usage in U.S. Consumer Battery Production" (1986), box 4, Legislative Commission on Waste Management, Administrative Records, Minnesota State Archives.

111 Marc Boolish (Energizer Brands), interview by author, August 27, 2013.

112 Mark Boolish, "Maximizing Battery Recycling in the United States," CPSC Producer Responsibility Workshop: Exploring Solutions for Household Batteries, San Francisco, May 12, 2011.

113 Samantha MacBride, *Recycling Reconsidered: The Present Failure and Future Promise of Environmental Action in the United States* (Cambridge, MA: MIT Press, 2012), chap. 3.

114 Terry Telzrow, "Greener Batteries Cooperation between Industry and the EPA," *IEEE Aerospace and Electronic Systems Magazine* 10, no. 5 (May 1995): 35–36.

115 Ray Balfour, "Written Comments Submitted during 60 Day Comment Period for CA Universal Waste Rule," DTSC Control Number R-97-08, 2001.

116 Gray Davis, "Veto Statement: SB2146," September 25, 2000, http://leginfo.public .ca.gov/pub/99-00/bill/sen/sb_2101-2150/sb_2146_vt_20000925.html.

117 California Environmental Protection Agency, "New Waste Disposal Rules Apply to Batteries, Fluorescent Lights, Four Year Exemption for Households, Small Businesses to Expire," January 23, 2006 (online press release no longer available).

118 County of San Diego, Department of Environmental Health, "Written Comments Submitted during 60 Day Comment Period for CA Universal Waste Rule," DTSC Control Number R-97-08, 2001.

119 "Recycle Any Battery or Electronic Device—The Big Green Box," October 1, 2003, https://web.archive.org/web/20031001132406/http://www.biggreenbox.com /StoreFront.bok.

120 Todd Coy (executive vice president of Kinsbursky Brothers International), interview by author, July 22, 2014; Paul Johnson (executive director of environmental affairs at Kinsbursky Brothers International), interview by author, July 22, 2014.

121 Olivetti, Gregory, and Kirchain, *Life Cycle Impacts of Alkaline Batteries with a Focus on End-of-Life*, 2011.

122 Balfour, "Household Battery Disposal: An Overview," in *First International Seminar on Battery Waste Management* (1989).

123 Olivetti, Gregory, and Kirchain, *Life Cycle Impacts of Alkaline Batteries with a Focus on End-of-Life*, 56–64.

124 Boolish, interview by author.

125 Khush Marolia (Global Product Stewardship at Procter & Gamble), interview by author, June 26, 2013.

126 2011 Battery Summit, Briefing Paper Factbase, Dallas, Texas, April 5–6, 2011.

127 Boolish, interview by author.

128 Corporation for Battery Recycling, "Corporation for Battery Recycling (CBR) Officially Releases RFP, Seeking Stewardship Organization to Oversee Voluntary National Household Battery Recycling Program," PR newswire, July 10, 2012.

129 Catherine Kavanaugh, "Environmental Groups Press Rayovac for Battery Recycling," *Waste and Recycling News*, August 1, 2013.

130 Robin Schneider (executive director, Texas Campaign for the Environment), interview by author, July 3, 2014.

131 Khush Marolia, interview by author.

132 Author's notes from presentations and discussion at National Battery Stewardship Meeting, Hartford, CT, June 11–12, 2014.

133 An Act Relating to Establishing a Product Stewardship Program for Primary Batteries, Act 139, Vermont State Legislature (2014).

134 Olivetti, Gregory, and Kirchain, *Life Cycle Impacts of Alkaline Batteries with a Focus on End-of-Life*, 23.

135 An Act Relating to Establishing a Product Stewardship Program for Primary Batteries, Act 139, Vermont State Legislature (2014).

136 Call2Recycle, Inc., *Vermont Primary Battery Stewardship Plan* (Atlanta, GA: Call2Recycle, November 20, 2020).

137 MacBride, *Recycling Reconsidered*.

138 "Burgess Laboratories," *Waste Trade Journal* 24, no. 23 (April 6, 1918): 18.

139 Chase Ezell, "Energizer EcoAdvanced Is the World's First AA Battery Made with Recycled Batteries," *Earth911*, February 3, 2015.

140 Energizer Holdings Inc, *Annual Report*, 2015, 30.

141 "EcoAdvanced," Energizer, January 2015, www.energizer.ca/docs/default-source /pdf/energizer-ecoadvanced-infographic-final.pdf.

142 Advertising Self-Regulatory Council, "Environmental Digest" (New York: National Advertising Division, August 2018).

143 Lee Presser, "A Conversation with Marc Boolish—Energizer Director of Technology," August 11, 2015, www.youtube.com/watch?v=vUVDpz9iAw8&frags= pl%2Cwn.

144 The changes in Energizer's advertising strategy can be tracked using energizer. com/batteries/battery-comparison-chart webpages archived at archive.org.

145 Giovanni Dolci et al., "Life Cycle Assessment of Consumption Choices: A Comparison between Disposable and Rechargeable Household Batteries," *International Journal of Life Cycle Assessment* 21 no. 12 (December 2016): 1691–705.

146 Omega, "Dry Batteries and How to Make Them," *Scientific American*, May 10, 1913, 435.

147 LeCain, *Mass Destruction*, 207–8.

3. LITHIUM-ION BATTERIES, THE SMARTPHONE, AND A WIRELESS REVOLUTION

1 Berit Larsson et al., "Lessons from the First Patient with an Implanted Pacemaker," *Pacing and Clinical Electrophysiology* 26, no. 1 (2003): 114–24.

2 G. Frank O. Tyers and Robert R. Brownlee, "Current Status of Pacemaker Power Sources," *Annals of Thoracic Surgery* 25, no. 6 (June 1978): 571–87.

3 Kurt Kelty, EV Fest, Benchmark Intelligence, online conference, May 29, 2020.

4 Bernadette Bensaude-Vincent and Arne Hessenbruch, "Interview with Michael Stanley Whittingham" (Pasadena: California Institute of Technology, History of Recent Science and & Technology, October 30, 2000).

5 M. S. Whittingham, "Electrical Energy Storage and Intercalation Chemistry," *Science* 192, no. 4244 (June 11, 1976): 1126–27.

6 Naomi Oreskes and Erik M. Conway, *Merchants of Doubt: How a Handful of Scientists Obscured the Truth on Issues from Tobacco Smoke to Global Warming* (New York: Bloomsbury, 2011).

7 M. Stanley Whittingham, "History, Evolution, and Future Status of Energy Storage," Special Centennial Issue, *Proceedings of the IEEE* 100 (May 2012): 1518–34.

8 Exxon, "America Wants a Big Car with a Small Energy Appetite" (Florham Park, NJ: Exxon, November 1978); Neela Banerjee, "For Exxon, Hybrid Car Technology Was Another Road Not Taken." *InsideClimate News*, October 5, 2016.

9 Metaphor drawn from Drew Baglino, Tesla Battery Day Presentation, September 22, 2020, www.tesla.com/2020shareholdermeeting.

10 Akira Yoshino, "Nobel Lecture," December 8, 2019, youtu.be/Q6pOFo9Cidw.

11 Akira Yoshino, "The Birth of the Lithium-Ion Battery," *Angewandte Chemie International Edition* 51, no. 24 (June 11, 2012): 5798–800.

12 Martin Winter, Brian Barnett, and Kang Xu, "Before Li Ion Batteries," *Chemical Reviews* 118, no. 23 (December 12, 2018): 11433–56.

13 "A Dream Comes True," Sony Corporation, www.sony.net/SonyInfo/Corporate Info/History/SonyHistory/2-13.html#block3.

14 Winter, Barnett, and Xu, "Before Li Ion Batteries," 11433–56.

15 Charles McCoy, "Valence's Pioneer Batteries Get Juice of a Big Motorola Order," *Wall Street Journal*, December 9, 1992.

16 Olof Ramström, "Interview about the 2019 Nobel Prize in Chemistry," The Nobel Prize, https://youtu.be/1bV8pmH7euA.

17 "Macintosh PowerBook 100: Technical Specifications," Apple, 1991, https://support .apple.com/kb/sp141?locale=en_US.

18 Fred Schlachter, "No Moore's Law for Batteries," *Proceedings of the National Academy of Sciences* 110, no. 14 (April 2, 2013): 5273.

19 This estimate is based on a comparison of a 1991 Cray Y-MP C90 supercomputer (16 gigaflop peak performance, 493.5 kW power consumption, 6136 kilograms, and $57 million adjusted for inflation) and an iPhone 11 powered by an A13 microprocessor (155 gigaflops, 5 watts, 194 grams, and $1,000). The best-selling 1991 Honda Accord cost $38,500 adjusted for inflation, had a top speed of 126 mph, and got 24 mpg. This is an imperfect comparison, but it helps illustrate the advances in the performance, energy efficiency, and cost of microprocessors.

20 Ralph J. Brodd, "Comments on the History of Lithium-Ion Batteries," 2002, www .electrochem.org/dl/ma/201/pdfs/0259.pdf.

21 David Linden and Thomas B. Reddy, *Handbook of Batteries* (New York: McGraw-Hill, 2002), 35.7.

22 Haresh Kamath, *Lithium Ion Batteries for Electric Transportation: Costs and Markets* (Electric Power & Research Institute, September 22, 2009), 7.

23 Ralph Brodd, "Recent Developments in Batteries for Portable Consumer Applications," *Interface*, Fall 1999, 20–23; Christopher Pillot. "The Worldwide Battery Market 2011–2025" (presentation, Batteries 2012, Nice, France, October 2012).

24 Lois Beckett, "By the Numbers: Life and Death at Foxconn," ProPublica, January 27, 2012.

25 Karin Valentine, "Class of Stellar Explosions Found to Be Galactic Producers of Lithium," *ASU Now*, June 1, 2020.

26 Ethan Siegel, "Lithium Mystery Solved: It's Exploding Stars, Not the Big Bang or Cosmic Rays," *Forbes*, June 3, 2020.

27 A. V. Tkachev, D. V. Rundqvist, and N. A. Vishnevskaya, "The Main Features of Lithium Metallogeny in Geological Time," *Doklady Earth Sciences* 484, no. 1 (January 1, 2019): 32–36.

28 "The Lithium Blues—Or How America Triggered an Out-of-Control Nuke," Medium, March 17, 2014.

29 Walter A. Brown, *Lithium: A Doctor, a Drug, and a Breakthrough* (New York: Liveright, 2019).

30 George E. Ericksen, "Lithium Resources of Salars in the Central Andes," in *Lithium Resources and Requirements by the Year 2000*, ed. James D. Vine (Washington, DC: USGS, 1976).

31 Patricio Garcia Méndez, *The Reinvention of the Saltpeter Industry* (Santiago, Chile: SQM, 2018); Sarah Rovang, "Incomplete Remains: Interpreting Mining Company Towns in Chile," Society of Architectural Historians, February 1, 2019.

32 Raul Ferro, "Soquimich Funds Expansions," *Chemical Week*, October 14, 1992.

33 George E. Ericksen, "Lithium Resources of Salars in the Central Andes," in *Lithium Resources and Requirements by the Year 2000*, ed. James D. Vine (Washington, DC: USGS, 1976), 66.

34 Roberto Rondanelli, Alejandra Molina, and Mark Falvey, "The Atacama Surface Solar Maximum," *Bulletin of the American Meteorological Society* 96, no. 3 (March 1, 2015): 405–18.

35 Remco Perotti and Manlio F. Coviello, *Governance of Strategic Minerals in Latin America: The Case of Lithium* (Santiago, Chile: Economic Commission for Latin America and the Caribbean, 2015).

36 Garcia Méndez, *The Reinvention of the Saltpeter Industry* (2018); Arlene Ebensperger et al., "The Lithium Industry: Its Recent Evolution and Future Prospects," *Resources Policy* 30, no. 3 (September 2005): 218–31; Haresh Kamath, *Lithium Ion Batteries for Electric Transportation: Costs and Markets* (Electric Power Research Institute, September 22, 2009), 21.

37 Hanjiro Ambrose and Alissa Kendall, "Understanding the Future of Lithium: Part 2, Temporally and Spatially Resolved Life-cycle Assessment Modeling," *Journal of Industrial Ecology* 24, no. 1 (February 2020): 90–100, https://doi.org/10.1111/jiec.12942.

38 SQM, *Sustainability Report* (2011), 85.

39 Cristian Fernando Olmos Herrera, "Hydrosocial Territories in the Atacama Desert: An Ethnographic Analysis of Changing Water Practices in Toconao, Chile" (Ph.D. diss., University College London, 2019), 218.

40 SQM, *Sustainability Report* (2011).

41 SQM, *Sustainability Report* (2011).

42 Gabriela Calderon, Marco Garrido, and Edmundo Acevedo, "Prosopis Tamarugo Phil.: A Native Tree from the Atacama Desert Groundwater Table Depth Thresholds for Conservation," *Revista Chilena de Historia Natural* 88, no. 1 (December 2, 2015): 18.

43 "Lithium Extraction for E-Mobility Robs Chilean Communities of Water," DW.com, January 23, 2020.

44 Lithium makes up 18.7 percent of the weight of lithium carbonate, so global trade in lithium carbonate was roughly one hundred thousand metric tons.

45 Statistics drawn from *2005 Minerals Yearbook* (Washington, DC: USGS, 2007).

46 Ellen Airhart, "Alternatives to Cobalt, the Blood Diamond of Batteries," *Wired*, June 7, 2018.

47 Nicholas D. Kristof, "Death by Gadget in Congo," *New York Times*, June 26, 2010.

48 Kim B. Shedd, "Cobalt," in *2005 Minerals Yearbook* (Washington, DC: USGS, 2007), 19.18.

49 P. A. Cox, *Transition Metal Oxides: An Introduction to Their Electronic Structure and Properties* (New York: Oxford University Press, 2010).

50 Edgar Peek, Torjus Akre, and Edouard Asselin, "Technical and Business Considerations of Cobalt Hydrometallurgy," *JOM* 61, no. 10 (October 2009): 43–53.

51 Marie Chêne, *Overview of Corruption and Anti-Corruption in the Democratic Republic of Congo* (Transparency International, October 8, 2010).

52 Adam Hochschild, *King Leopold's Ghost: A Story of Greed, Terror, and Heroism in Colonial Africa* (New York: Houghton Mifflin, 1998).

53 Hubert André-Dumont, McGuire Woods LLP, "Current DR Congo Mining Code under Revision," Lexology, 2002.

54 Nicolas Tsurukawa, Siddharth Prakash, and Andreas Manhart, "Social Impacts of Artisanal Cobalt Mining in Katanga, Democratic Republic of Congo," *Öko-Institut e.V.* (Freiburg, Germany, November 2011), 32.

55 Amnesty International and AfreWatch, "'This Is What We Die For': Human Rights Abuses in the Democratic Republic of the Congo Power the Global Trade in Cobalt" (London: Amnesty International, January 16, 2016), 19–20.

56 Tsurukawa, Prakash, and Manhart "Social Impacts of Artisanal Cobalt Mining in Katanga, Democratic Republic of Congo," 3.

57 Tsurukawa, Prakash, and Manhart "Social Impacts of Artisanal Cobalt Mining in Katanga, Democratic Republic of Congo," 32.

58 Collingsworth et al. v. Apple, Inc. et al., United States District Court for the District of Columbia, December 15, 2019.

59 "Lundin Family Confirms Big Zaire Copper Project," *National Post* (Toronto), July 24, 1996.

60 "Tenke Mining to Give $50 Million to Rebels," Associated Press, May 9, 1997.

61 Rod Nutt, "Congo Upheaval Causes Tenke to Halt Mine Project," *Vancouver Sun,* February 24, 1999; Chris McGreal, "A Country Pays as Foreigners Fight for the Spoils of War," *Guardian,* January 16, 2001; Claudia Carpenter, "Phelps Dodge Buys Congo Mining Rights," *Arizona Daily Star,* November 3, 2005.

62 Paul Luke, "Tenke Fungurume Gets Go-Ahead," *Times Colonist* (Victoria, British Columbia), December 8, 2006.

63 Johanna Jansson, *CSR Practice in the DRC's Mining Sector by Chinese Firms* (Africa Institute of South Africa [AISA], 2010), 1.

64 Benjamin Rubbers, "Mining Boom, Labour Market Segmentation and Social Inequality in the Congolese Copperbelt," *Development and Change* (2019), 17.

65 Prince Kumwamba and Anne-Sophie Simpere, "Soul Mining: The EIB's Role in the Tenke-Fungureme Mine, DRC" (CEE Bankwatch Network, August 2008), 17.

66 Rubbers, "Mining Boom, Labour Market Segmentation and Social Inequality in the Congolese Copperbelt," 18–19.

67 Kumwamba and Simpere, "Soul Mining," 17; Liezel Hill, "Freeport-McMoRan Meets DRC Protesters after 'Misunderstanding,'" *Mining Weekly,* January 15, 2008.

68 Kumwamba and Simpere, "Soul Mining," 17.

69 Linda Gaines and Roy Cuenca, "Costs of Lithium-Ion Batteries for Vehicles" (Argonne, IL: Argonne National Laboratory, May 2000).

70 "LG Chem to Hike Lithium-Ion Battery Price," *Metals Week,* March 15, 2004.

71 Célestin Lubaba Nkulu Banza et al., "High Human Exposure to Cobalt and Other Metals in Katanga, a Mining Area of the Democratic Republic of Congo," *Environmental Research* 109, no. 6 (August 2009): 745–52.

72 Keith Bradsher, "China Tightening Control of Rare Earth Minerals," *New York Times,* August 31, 2009.

73 Julie Michelle Klinger, *Rare Earth Frontiers: From Terrestrial Subsoils to Lunar Landscapes* (Ithaca, NY: Cornell University Press, 2018).

74 Klinger, *Rare Earth Frontiers.*

75 Bahar Moradi and Gerardine G. Botte, "Recycling of Graphite Anodes for the Next Generation of Lithium Ion Batteries," *Journal of Applied Electrochemistry* 46, no. 2 (February 2016): 123–48; Allah D. Jara et al., "Purification, Application and Current Market Trend of Natural Graphite: A Review," *International Journal of Mining Science and Technology* 29, no. 5 (September 1, 2019): 671–89.

76 Peter Whoriskey, "The Batteries in Your Favorite Devices Are Literally Covering Chinese Villages in Black Soot," *Washington Post*, October 2, 2016.

77 Liu Xin Xu Xindong, "Hundred Acres of Wheat Fields Were Polluted by Graphite and Other Dust, and Two Sewage Companies Were Investigated and Punished," *Peninsula City News*, June 16, 2011, http://green.sohu.com/20110616/n310322956 .shtml. Google Translate Chinese (Simplified) to English.

78 "Pollution," General Office of the People's Government of Shandong Province, June 15, 2011, http://gb.shandong.gov.cn/art/2013/4/2/art_100084_7203.html. Google Translate Chinese (Simplified) to English.

79 Ning Cui et al., "Geological Characteristics and Analysis of Known and Undiscovered Graphite Resources of China," *Ore Geology Reviews* 91 (December 2017): 1120.

80 "Pingdu No. 1 Graphite Factory Invaded and Polluted the Environment and Was Suspended for Rectification," Qingdao Network TV Station, August 2, 2012. Google Translate Chinese (Simplified) to English.

81 Whoriskey, "The Batteries in Your Favorite Devices Are Literally Covering Chinese Villages in Black Soot."

82 Pingdu Government Network, "Pingdu City Highlights the Key Points of Remediation and Effectively Fights the Tough Battle of Pollution Remediation in the Graphite Industry," August 26, 2014. www.pingdu.gov.cn/n2 /n1293/n1294/n1297/150522163950505434.html. Google Translate Chinese (Simplified) to English.

83 Simon Moores, "Graphite Red Alert," *Industrial Minerals*, March 2012.

84 "The 13th Five-Year Plan for Economic and Social Development of the People's Republic of China (2016–2020)" (Beijing, China: Central Committee of the Communist Party of China, 2015), https://en.ndrc.gov.cn/policyrelease_8233 /201612/P020191101482242850325.pdf.

85 "南海资讯," September 30, 2018, www.whnh.gov.cn/art/2018/9/30/art_40965 _1444864.html. Google Translate Chinese (Simplified) to English.

86 John A. Kukowki and William R. Boulton, "Electronics Manufacturing and Assembly in Japan," in *JTEC Panel Report on Electronic Manufacturing and Packaging in Japan* (Baltimore: International Technology Research Institute, February 1995), 97–103.

87 R. J. Brodd, *Factors Affecting U.S. Production Decisions: Why Are There No Volume Lithium-Ion Battery Manufacturers in the United States?* (Gaithersburg, MD: National Institute of Standards and Technology, June 2005).

88 P. G. Patil, *Developments in Lithium-Ion Battery Technology in the People's Republic of China* (Argonne, IL: Argonne National Laboratory, February 28, 2008), 4.

89 Yi Zhang, "The Story of Shenzhen: Its Economic, Social, and Environmental Transformation" (Nairobi, Kenya: United Nations Human Settlement Programme, 2019).

90 Taylor Lynch Ogan and Xiangming Chen, "The Rise of Shenzhen and BYD—How a Chinese Corporate Pioneer Is Leading Greener and More Sustainable Urban Transportation and Development," *European Financial Review*, February-March 2016, 9; Robert S. Huckman and Alan D. MacCormack, *BYD Company, Ltd.*, case study (Cambridge, MA: Harvard Business School, September 15, 2009).

91 "'Battery King' Winning Secret," July 15, 2002, https://zhuanlan.zhihu.com /p/141100558. Google Translate Chinese (Simplified) to English.

92 Wang Chuanfu, "王传福: 作为民营企业比迪亚的优势_搜狐 | Wang Chuanfu: The Advantages of Bidia as a Private Enterprise," August 7, 2004, https://it .sohu.com/20040807/n221413843.shtml. Google Translate Chinese (Simplified) to English.

93 Huckman and MacCormack, *BYD Company, Ltd.*

94 "'Battery King' Winning Secret."

95 Li Qiang, *BYD Company Investigative Report* (New York: China Labor Watch, June 2011).

96 Johanna Jansson, *CSR Practice in the DRC's Mining Sector by Chinese Firms* (Pretoria: Africa Institute of South Africa, 2010).

97 Andoni Maiza-Larrarte and Gloria Claudio-Quiroga, "The Impact of Sicomines on Development in the Democratic Republic of Congo," *International Affairs* 95, no. 2 (March 1, 2019): 434.

98 "Freeport Agrees to Sell Its Tenke Copper Mine," *Economist*, May 18, 2016.

99 Brian W. Jaskula, "Lithium," in *2011 Minerals Yearbook* (Washington, DC: USGS, 2013), 44.6.

100 Mathew Hocking et al., *Industry: Lithium 101*, Deutsche Bank, May 9, 2016, www .slideshare.net/stockshaman/deutsche-bank-lithium-report-may-2016.

101 Pillot, "The Worldwide Battery Market 2011–2025."

102 "'Exploding' Dell Laptop Destroys Truck, Imperils Outsdoorsmen," *Consumer Affairs*, August 3, 2006.

103 K. Ozawa, "Lithium-Ion Rechargeable Batteries with LiCoO2 and Carbon Electrodes: The LiCoO2/C System," *Solid State Ionics* 69, no. 3–4 (August 1994): 212–21.

104 Matthew N. Eisler, "Exploding the Black Box: Personal Computing, the Notebook Battery Crisis, and Postindustrial Systems Thinking," *Technology and Culture* 58, no. 2 (2017): 368–91.

105 Matthew N. Eisler, "The History of Lithium Ion Batteries Is Explosive," *Slate*, September 13, 2016.

106 Matthew E. Rossheim et al., "Electronic Cigarette Explosion and Burn Injuries, US Emergency Departments 2015–2017," *Tobacco Control* 28, no. 4 (July 1, 2019): 472–74.

107 Quotes from Nobel Prize Lectures, December 10, 2019, www.nobelprize.org
/prizes/chemistry/2019/.

108 "The Battery Megafactories Are Coming," *Benchmark Minerals*, April 3, 2016.

4. ELECTRIC CARS, TESLA, AND A ZERO-EMISSIONS FUTURE

1 Vehicle Technologies Office, "Fact #921: Japan Produced the Most Automotive
Lithium-Ion Batteries by Capacity in 2014," Office of Energy Efficiency & Renewable Energy, April 18, 2016.

2 Jerry Hirsch, "Tesla Plans to Raise Almost $2 Billion to Build Battery Factory," *Los
Angeles Times*, February 26, 2014.

3 Elon Musk, "The Secret Tesla Motors Master Plan (Just between You and Me),"
Tesla, August 2, 2006, www.tesla.com/blog/secret-tesla-motors-master-plan
-just-between-you-and-me.

4 "Gigafactory," Tesla, February 26, 2014, www.tesla.com/blog/gigafactory.

5 Jack Kaskey and Simon Casey, "Tesla to Use North American Material amid
Pollution Worry," *Bloomberg*, March 31, 2014.

6 Tesla, "Gigafactory Process Flow" presentation slides, Tesla, February 26,
2014.

7 Amory B. Lovins et al., *The Coming Light-Vehicle Revolution*, pdf (Snowmass,
CO: Rocky Mountain Institute, March 31, 1993), 34; Amory B. Lovins and
Hunter L. Lovins, "Reinventing the Wheels," *Atlantic Monthly*, January 1995, 13.

8 Gustavo Oscar Collantes, "The California Zero-Emission Vehicle Mandate: A Study
of the Policy Process, 1990–2004" (Ph.D. diss., University of California, Davis,
2004), 18.

9 Kristine Stiven Breese, "The Mandate Is Clear, Dirty Air Must Go," *Californian*,
March 8, 1991.

10 National Research Council, *Effectiveness of the United States Advanced Battery
Consortium as a Government-Industry Partnership* (Washington, DC: National
Academies Press, 1998).

11 Owen Edwards, "The Death of the EV-1," *Smithsonian Magazine*, June 2006.

12 Toyota, "Eat My Voltage," advertisement, *Popular Science*, September 2000, 9.
Advertised fuel economy in 2000.

13 *Consumer Reports*, Auto Issue, April 2005.

14 Timothy Cain, "Top 10 Best-Selling Cars in America—2007 Year End," 2008,
Goodcarbadcar, www.goodcarbadcar.net/usa-10-best-selling-cars-2007-year
-end/.

15 Christopher Trout, "Toyota Sells One Million Prii in US Alone," *Engadget*, April 7,
2011.

16 Aaron Robinson, "2001 Toyota Prius: Perhaps the First Car That Runs on Guilt,"
Car and Driver, March 1, 2001.

17 James Morton Turner and Andrew C. Isenberg, *The Republican Reversal: Conservatives and the Environment from Nixon to Trump* (Cambridge, MA: Harvard
University Press, 2018), chap. 4.

18 Austan D. Goolsbee and Alan B. Krueger, "A Retrospective Look at Rescuing and Restructuring General Motors and Chrysler" (working paper, Princeton University, Industrial Relations Section, February 19, 2015).

19 Jody Freeman, "The Obama Administration's National Auto Policy: Lessons from the 'Car Deal,'" *Harvard Environmental Law Review* 35 (2011): 32.

20 Joseph E. Aldy, "Policy Monitor a Preliminary Assessment of the American Recovery and Reinvestment Act's Clean Energy Package," *Review of Environmental Economics and Policy* 7, no. 1 (Winter 2013); Bill Canis and Brent Yacobucci, "The Advanced Technology Vehicles Manufacturing Loan Program: Status and Issues," Congressional Research Service, January 15, 2015.

21 Barack Obama, "Remarks by the President on the American Automotive Industry," The White House, March 3, 2009.

22 Barack Obama, "Remarks at a Town Hall Meeting and a Question-and-Answer Session in Henderson, Nevada," American Presidency Project, February 19, 2010.

23 "A Retrospective Assessment of Clean Energy Investments in the Recovery Act," The White House, February 2016.

24 Seth Fletcher, *Bottled Lightning: Superbatteries, Electric Cars, and the New Lithium Economy* (New York: Hill and Wang, 2011).

25 Barack Obama, "Remarks by the President in Phone Call to Recovery Act Advanced Battery Grant Recipient, A123 Systems in Livonia, MI," The White House, September 13, 2010.

26 Mark Gillies, "Henrik Fisker: The Man behind the 2010 Fisker Karma," *Car and Driver*, June 23, 2009.

27 Elizabeth Alexander, "'You've Got to Believe'—Building the Cars of the Future in America," The White House, October 27, 2009.

28 "AVTA: ARRA EV Project Overview Reports," Department of Energy, www .energy.gov/eere/vehicles/downloads/avta-arra-ev-project-overview-reports.

29 Bjorn Lomborg, "Green Cars Have a Dirty Little Secret," *Wall Street Journal*, March 11, 2013.

30 Don Anair and Amine Mahmassani, *State of Charge* (Cambridge, MA: UCS Publications, 2012), 10.

31 Lomborg, "Green Cars Have a Dirty Little Secret."

32 "Greenhouse Gas Inventory Data Explorer," Environmental Protection Agency, https://cfpub.epa.gov/ghgdata/inventoryexplorer.

33 Environmental Protection Agency, "Revisions and Additions to Motor Vehicle Fuel Economy Label," *Federal Register* 76, no. 129 (July 6, 2011): 39478–586.

34 Dominic A. Notter et al., "Contribution of Li-Ion Batteries to the Environmental Impact of Electric Vehicles," *Environmental Science & Technology* 44, no. 17 (September 1, 2010): 6550–56; Rolf Frischknecht and Karin Flury, "Life Cycle Assessment of Electric Mobility: Answers and Challenges—Zurich, April 6, 2011," *International Journal of Life Cycle Assessment* 16, no. 7 (August 2011): 691–95; Jeremy J. Michalek et al., "Valuation of Plug-in Vehicle Life-Cycle Air

Emissions and Oil Displacement Benefits," *Proceedings of the National Academy of Sciences* 108, no. 40 (October 4, 2011): 16554–58.

35 Estimate based on Troy R. Hawkins et al., "Comparative Environmental Life Cycle Assessment of Conventional and Electric Vehicles," *Journal of Industrial Ecology* 17, no. 1 (February 2013): 53–64.

36 Rachael Nealer et al., *Cleaner Cars from Cradle to Grave: How Electric Cars Beat Gasoline Cars on Lifetime Global Warming Emissions* (Cambridge, MA: Union of Concerned Scientists, 2015).

37 J. B. Dunn et al., "The Significance of Li-Ion Batteries in Electric Vehicle Life-Cycle Energy and Emissions and Recycling's Role in Its Reduction," *Energy & Environmental Science* 8 (2015): 160.

38 Dunn et al., "The Significance of Li-Ion Batteries."

39 Dunn et al., "The Significance of Li-Ion Batteries."

40 Benchmark Mineral Intelligence, "Elon Musk: Our Lithium Ion Batteries Should Be Called Nickel-Graphite," *Benchmark Mineral Intelligence*, June 5, 2016.

41 Naoki Nitta et al., "Li-Ion Battery Materials: Present and Future," *Materials Today* 18, no. 5 (June 2015): 252–64.

42 In October 2020, Tesla began using lithium iron phosphate (LFP) battery cells in some Chinese-manufactured vehicles.

43 Wangda Li, Evan M. Erickson, and Arumugam Manthiram, "High-Nickel Layered Oxide Cathodes for Lithium-Based Automotive Batteries," *Nature Energy* 5, no. 1 (January 2020): 26.

44 Nicolò Campagnol et al., *The Future of Nickel: A Class Act* (McKinsey & Company, November 2017).

45 Campagnol et al., "The Future of Nickel."

46 Kunio Kaiho, David S. Jones, and Li Tian, "Pulsed Volcanic Combustion Events Coincident with the End-Permian Terrestrial Disturbance and the Following Global Crisis," *Geology* 49, no. 3 (2021): 289–93.

47 Alexander Yakubchuk and Anatoly Nikishin, "Noril'sk–Talnakh Cu–Ni–PGE Deposits: A Revised Tectonic Model," *Mineralium Deposita* 39, no. 2 (March 1, 2004): 125–42; Jun Shen et al., "Evidence for a Prolonged Permian–Triassic Extinction Interval from Global Marine Mercury Records," *Nature Communications* 10, no. 1 (April 5, 2019): 1563.

48 Marlene Laruelle and Sophie Hohmann, "Biography of a Polar City: Population Flows and Urban Identity in Norilsk," *Polar Geography* 40, no. 4 (October 2, 2017): 310.

49 Laruelle and Hohmann, "Biography of a Polar City," 309.

50 Laruelle and Hohmann, "Biography of a Polar City," 317.

51 Blacksmith Institute, *The World's Worst Polluted Places: The Top Ten* (New York: Blacksmith Institute, October 2006).

52 O. N. Zubareva et al., "Zoning of Landscapes Exposed to Technogenic Emissions from the Norilsk Mining and Smelting Works," *Russian Journal of Ecology* 34, no. 6 (2003): 6.

53 Larisa Bronder et al., *Norilsk Nickel: The Soviet Legacy of Industrial Pollution* (Murmansk, Russia: Bellona Foundation, 2011).

54 "Открытое Письмо Генеральному Директору ГМК 'Норильский Никель,'" UC Rusal, August 12, 2008, http://krsk.sibnovosti.ru/society/56359-otkrytoe -pismo-generalnomu-direktoru-gmk-norilskiy-nikel. Google Translate, Russian to English.

55 Henry Foy, "Polluter Norilsk Nickel Forced to Clean up Its Act," *Financial Times*, April 30, 2018.

56 Environmental Protection Agency, "Cleaner Production Project at Polar Branch of Norilsk Nickel Company in Arctic," https://archive.epa.gov/international /regions/web/html/russiapast.html.

57 Council on Ethics, *Recommendation of 16 February 2009 re: Norilsk Nickel*, (Norway: Government Pension Fund Global, 2009).

58 Nornickel, *2018 Sustainability Report: The New Nornickel: Strategy in Action*, May 27, 2019.

59 Anton Troianovski, "On 'Island' in Russian Arctic, Arrival of Fast Internet Shakes Political Calm," *New York Times*, October 20, 2019.

60 Gennady Schukin, "An Appeal of Aborigen-Forum Network to Elon Musk, the Head of the Tesla Company," iRussia, August 6, 2020.

61 Johannes Rohr, *Indigenous Peoples in the Russian Federation* (Copenhagen, Denmark: International Work Group for Indigenous Affairs, 2014), 41–42.

62 Nornickel, *2018 Sustainability Report.*

63 "Norilsk Nickel Commits over $25M to Support Indigenous Population in Russian Far North," press release, MMC Norilsk Nickel, September 30, 2020.

64 Brian Eckhouse and Yvonne Yue Li, "Musk Promises 'Giant Contract' for Efficiently Mined Nickel," *Bloomberg*, July 22, 2020.

65 Kevin Murphy, "Battery-Grade Nickel Supply Will Suffer as Major Nickel Discoveries Slump," blog post, S&P Global Market Intelligence, June 12, 2020.

66 US Geological Survey, Mineral Commodity Summaries, Lithium, 2010–2019, www.usgs.gov/centers/nmic/lithium-statistics-and-information.

67 J.-M. Tarascon, "Is Lithium the New Gold?," *Nature Chemistry* 2 (June 2010): 510.

68 Javiera Barandiarán, "Lithium and Development Imaginaries in Chile, Argentina and Bolivia," *World Development* 113 (January 2019): 381–91.

69 Lawrence Wright, "Lithium Dreams: Can Bolivia Become the Saudi Arabia of the Electric-Car Era?," *New Yorker*, March 22, 2010.

70 Comisión Nacional del Litio, "Litio: Una Fuente de Energiá Una Oportunidad Para Chile," Informe Final (Santiago, Chile: Ministerio de Mineria, 2015).

71 US Geological Survey, Mineral Commodity Summaries, Lithium, 2010.

72 Marcela Valente, "Native People in Argentina Demand a Say in Lithium Mining," *Inter Press Service,* posted by Global Issues, March 29, 2012.

73 Pía Marchegiani, Jasmin Höglund Hellgren, and Leandro Gómez, *Lithium Extraction in Argentina: A Case Study on the Social and Environmental Impacts* (Fundación Ambiente y Recursos Naturales, 2019).

74 USGS, Mineral Commodity Summaries, Lithium, 2010–2018.

75 David I. Groves, Mark E. Barley, and Julie M. Shepherd, "Overviews: Geology and Mineralisation of Western Australia," *ASEG Extended Abstracts*, no. 1 (December 1994): 1–28.

76 Peter Ker, "Lithium Superpower: How Australia's Penny-Dreadfuls Stole a March on the World," *Financial Review*, June 15, 2018; Laura Talens Peiró, Gara Villalba Méndez, and Robert U. Ayres, "Lithium: Sources, Production, Uses, and Recovery Outlook," *JOM* 65, no. 8 (August 1, 2013): 986–96.

77 Brian Fitch, Marina Yakovleva, and Scott Meiere, "Lithium Hydroxide Based Performance Improvements for Nickel Rich NCM Layered Cathode Material," *ECS Meeting Abstracts,* MA2016-02 (2016) 469.

78 Alex Grant et al., "The CO2 Impact of the 2020s' Battery Quality Lithium Hydroxide Supply Chain," blog post, Minviro, January 2020.

79 Stephen E. Kesler et al., "Global Lithium Resources: Relative Importance of Pegmatite, Brine and Other Deposits," *Ore Geology Reviews* 48 (2012): 55–69.

80 Extent of mining operations estimated using satellite imagery and measurements from Google Earth, accessed August 2020.

81 "Lithium Contract Win for NPJV," *Pick Online Magazine*, October 4, 2017; Jeremy Goff, "Njamal People Are on the Road Again," *National Indigenous Times*, January 12, 2018.

82 Ker, "Lithium Superpower."

83 *Green Energy Oversight: Examining the Department of Energy's Bad Bet on Fisker Automotive: Hearing Before the Subcommittee on Economic Growth, Job Creation, and Regulatory Affairs*, U.S. House of Representatives, 113th Cong., 1st Sess. (2013) (questions from Rep. Patrick T. McHenry, R-NC).

84 *Risky Business: The DOE Loan Guarantee Program, Hearing Before the Committee on Science, Space, and Technology,* U.S. House of Representatives, 115th Cong., 1st Sess. (Feb. 15, 2017) (testimony of Dan Reicher, Stanford University.)

85 Bill Canis and Brent D. Yacobucci, "The Advanced Technology Vehicles Manufacturing (ATVM) Loan Program: Status and Issues," Congressional Research Service, January 15, 2015, 17.

86 *Green Energy Oversight Hearing* (questions from Rep. Ron DeSantis, R-FL).

87 "Tesla Gets Loan Approval from US Department of Energy," Tesla, press release, April 20, 2010.

88 Scott Woolley, "Actually, the U.S. Government's Investment in Tesla Was a Disaster, and Cost Taxpayers at Least $1 Billion," *Slate Magazine*, May 29, 2013.

89 "Plug-In Electric Car Sales for 2015 Fall Slightly from 2014," Green Car Reports, January 19, 2016.

90 Turner and Isenberg, *The Republican Reversal*, conclusion.

91 Shawn Ou et al., *A Study of China's Explosive Growth in the Plug-in Electric Vehicle Market* (Oak Ridge, TN: Oak Ridge National Laboratory, 2017), 37.

92 Huckman and MacCormack, *BYD Company, Ltd.*, case study, 2019.

93 Hua Wang and Chris Kimble. "Betting on Chinese Electric Cars?—Analysing BYD's Capacity for Innovation," *International Journal of Automotive Technology and Management* 10, no. 1 (2010): 77.

94 Matthew Campbell and Ying Tian, "The World's Biggest Electric Vehicle Company Looks Nothing Like Tesla," *Bloomberg Businessweek*, April 16, 2019.

95 Lingzhi Jin et al., *Driving a Green Future: A Retrospective Review of China's Electric Vehicle Development and Outlook for the Future* (Washington, DC: International Council on Clean Transportation, 2020), 8.

96 Christopher Pillot, "The Rechargeable Battery Market and Main Trends 2018–2030," *Avicenne Energy* (May 24, 2019), 13.

97 Donna Lu, "Forget Tesla—China's BYD Is Driving the Electric Car Revolution," *New Scientist*, July 10, 2019.

98 *Impact of COVID-19 on Mineral Supply Chains, Hearing Before the Committee on Energy and Natural Resources,* U.S. Senate, 116th Cong., 2nd Sess. (June 24, 2020) (statement of Simon Moores, Benchmark Mineral Intelligence).

99 John D. Graham et al., "How China Beat the US in Electric Vehicles," *Issues in Science and Technology* (Winter 2021).

100 *The Outlook for Energy and Minerals Markets in the 116th Congress, Hearing Before the Committee on Energy and Natural Resources,* U.S. Senate, 116th Cong., 1st Sess. (February 5, 2019) (statement of Simon Moores, Benchmark Mineral Intelligence).

101 Shunsuke Tabeta, "China Plans to Phase out Conventional Gas-Burning Cars by 2035," Nikkei Asia, October 27, 2020.

102 "Norwegian EV Policy," Norsk elbilforening, https://elbil.no/english/norwegian-ev-policy/.

103 "New Mercedes," skit, *Saturday Night Live*, NBC, April 16, 2016.

104 Nissan, "2011 LEAF: Dismantling Guide," Pub. No. DG1E-1ZE0U1, 2012.

105 Kurt Kelty, "Mythbusters Part 3: Recycling Our Non-Toxic Battery Packs," Tesla, March 11, 2008.

106 Linda Gaines, "The Future of Automotive Lithium-Ion Battery Recycling: Charting a Sustainable Course," *Sustainable Materials and Technologies* 1–2 (December 2014): 2–7.

107 Xu et al., "Future Material Demand for Automotive Lithium-Based Batteries," fig. 4.

108 General Motors, "GM and ABB Show Chevy Volt Battery Reuse Application," 3BL Media, November 14, 2012; Jim Motavalli, "E.V. Batteries Can Have Second Lease on Life, Automakers Assert," *New York Times*, July 21, 2011.

109 Mona Moll, "World's Largest 2nd-Use Battery Storage Is Starting Up," Daimler, press release, September 13, 2016.

110 Tobias Elwert et al., "Current Developments and Challenges in the Recycling of Key Components of (Hybrid) Electric Vehicles," *Recycling* 1, no. 1 (October 22, 2015): 25–60.

111 Xiaohong Zheng et al. "A Mini-Review on Metal Recycling from Spent Lithium-Ion Batteries," *Engineering* 4, no. 3 (June 2018): 361–70.

112 Gavin Harper et al., "Recycling Lithium-Ion Batteries from Electric Vehicles," *Nature* 575, no. 7781 (November 2019): 82–83.

113 Hans Eric Melin, "Start-of-the-Art in Reuse and Recycling of Lithium-Ion Batteries—A Research Review" (London: Circular Energy Storage, 2019), 27–32.

114 Tim Higgins, "One of the Brains behind Tesla May Have a New Way to Make Electric Cars Cheaper," *Wall Street Journal*, August 29, 2020; "Redwood's Plan to Produce Sustainable Battery Materials," Redwood Materials, press release, September 14, 2021.

115 Tobias Elwert et al., "Recycling of Batteries from Electric Vehicles," in *Behaviour of Lithium-Ion Batteries in Electric Vehicles: Battery Health, Performance, Safety, and Cost*, ed. Gianfranco Pistoia and Boryann Liaw (Cham, Switzerland: Springer International, 2018), 289–321.

116 Committee on Industry, Research, and Energy, *Report on a Comprehensive European Approach to Energy Storage*, European Parliament, 2019.

117 "China Puts Responsibility for Battery Recycling on Makers of Electric Vehicles," Reuters, February 26, 2018.

118 Melin, "Start-of-the-Art in Reuse and Recycling of Lithium-Ion Batteries."

119 *Effectiveness of the United States Advanced Battery Consortium as a Government-Industry Partnership* (Washington, DC: National Academies Press, 1998).

120 McKinsey Sustainability & Resource Productivity, *Energy Efficiency: A Compelling Global Resource*, ed. Tom Kiely (2010), 55. These are estimates of pack-level battery costs.

121 Andrew Leyland, "Lithium Ion Battery Cell Prices Fall to $110/KWh," *Benchmark Mineral Intelligence*, December 1, 2020.

122 Alexander Bills, Shashank Sripad, and Venkat Viswanathan, "The Road to Electric Vehicles with Lower Sticker Prices than Gas Cars—Battery Costs Explained," *The Conversation*, July 27, 2020.

123 Fletcher, *Bottled Lightning* (2011).

124 Andrew Leyland, "Lithium Ion Battery Cell Prices Fall to $110/kWh," *Benchmark Mineral Intelligence*, December 1, 2020.

CONCLUSION

1 Alexandria Ocasio-Cortez, "A Message from the Future with Alexandria Ocasio-Cortez," *The Intercept*, April 17, 2019, YouTube video, 7:35.

2 Recognizing the Duty of the Federal Government to Create a Green New Deal, H.R. Res. 109, 116th Cong. (2019).

3 Jedediah Britton-Purdy, "The Green New Deal Is What Realistic Environmental Policy Looks Like," *New York Times*, February 16, 2019.

4 Naomi Klein, *This Changes Everything: Capitalism vs. The Climate* (New York: Simon and Schuster, 2014); Naomi Klein, *On Fire: The (Burning) Case for a Green New Deal* (New York: Simon and Schuster, 2020).

5 Eric Klinenberg, "The Great Green Hope," *New York Review*, April 23, 2020.

6 Kirsten Hund et al., *The Mineral Intensity of the Clean Energy Transition* (Washington, DC: World Bank Group, 2020), 11.

7 Calculations based on Hund et al., *The Mineral Intensity of the Clean Energy Transition*; and "Coal Information: Overview," International Energy Agency (July 2020), www.iea.org/reports/coal-information-overview.

8 James Morton Turner, *The Promise of Wilderness: American Environmental Politics since 1964* (Seattle: University of Washington Press, 2012).

9 Barry Commoner, *The Closing Circle: Nature, Man, and Technology* (New York: Random House, 1971).

10 Donald Worster, *The Wealth of Nature: Environmental History and the Ecological Imagination* (New York: Oxford University Press, 1993), 143.

11 E. F. Schumacher, *Small Is Beautiful: Economics as If People Mattered* (New York: Harper Perennial, 1973). On the historical importance of the appropriate technology movement, see Andrew G. Kirk, *Counterculture Green: The Whole Earth Catalog and American Environmentalism* (Lawrence: University Press of Kansas, 2007); Sarah Mittlefehldt, "From Appropriate Technology to the Clean Energy Economy: Renewable Energy and Environmental Politics since the 1970s," *Journal of Environmental Studies and Sciences* 8, no. 2 (June 2018): 212–19.

12 Jeremy Caradonna et al., "A Call to Look Past an Ecomodernist Manifesto: A Degrowth Critique," 2015, www.resilience.org/stories/2015-05-06/a-degrowth-response-to-an-ecomodernist-manifesto/.

13 Matt T. Huber, "Ecological Politics for the Working Class," *Catalyst* 3, no. 1 (Spring 2019).

14 Jesse H. Ausubel, "Can Technology Spare the Earth?," *American Scientist* 84, no. 2 (March 1996).

15 Aaron Greenfield and T. E. Graedel, "The Omnivorous Diet of Modern Technology," *Resources, Conservation and Recycling* 74 (May 2013): 1–7.

16 Alexander H. King, "Our Elemental Footprint," *Nature Materials* 18, no. 5 (May 2019): 408–9.

17 Greenfield and Graedel, "The Omnivorous Diet of Modern Technology," 1–7.

18 Bruno Latour, "An Attempt at a 'Compositionist Manifesto,'" *New Literary History* 41, no. 3 (2010): 471–90; Bruno Latour, "Love Your Monsters—Why We Must Care for Our Technologies As We Do Our Children," Breakthrough Institute, 2011, https://thebreakthrough.org/journal/issue-2/love-your-monsters.

19 John D. Graham et al., "How China Beat the US in Electric Vehicles," *Issues in Science and Technology* Winter (2021); Lingzhi Jin et al., "Driving a Green Future: A Retrospective Review of China's Electric Vehicle Development and Outlook for the Future," International Council on Clean Transportation, January 2021.

20 Letter Opposing the American Mineral Security Act from Environmental Groups, September 15, 2020.

21 A notable exception is Kate Aronoff et al., *A Planet to Win: Why We Need a Green New Deal* (London: Verso, 2019), chap. 4.

22 Jason Potts, Matthew Wenban-Smith, and Laura Turley, *State of Sustainability Initiatives Review: Standards and the Extractive Economy*, 2018, www.deslibris.ca/ID/10097867.

23 Initiative for Responsible Mining Assurance, *IRMA Standard for Responsible Mining* (Initiative for Responsible Mining Assurance, June 2018).

24 Initiative for Responsible Mining Assurance, "Chain of Custody: Standard for Responsibly Mined Minerals" [draft], October 15, 2020.

25 Mark J. Perry, "To Reduce China's Leverage, Rebuild America's Minerals Supply Chain," *The Hill,* June 26, 2018; "Senate Minerals Permitting Legislation Recognizes Vital Role of the Raw Materials Necessary to Our National and Economic Security," National Mining Association, blog, May 2, 2019; "LETTER: Western Governors Support American Minerals Security Act," Western Governors' Association, June 4, 2019, westgov.org/news/article/western-governors-support-american-minerals-security-act-to-improve-access-to-domestic-minerals-strengthen-economy.

26 American Mineral Security Act. S.1317. 116th Cong. (2019).

27 Letter Opposing the American Mineral Security Act from Environmental Groups, September 15, 2020, www.congress.gov/116/meeting/house/111008/documents/HHRG-116-IF18-20200916-SD003.pdf.

28 Analysis based on data drawn from the respective commodity analyses in the USGS, *Minerals Yearbooks* for 1990 to 2019. Production data is based on refined metal production (which is reportedly consistently) not primary ore mined (which is not reported for all commodities).

29 Based on mining and processing of nickel cobalt, graphite, lithium, and manganese, as reported by Simon Moores, Presentation, Benchmark Intelligence, Benchmark Summit, October 10, 2020.

30 Hardrock Leasing and Reclamation Act, H.R. 2579, 116th Cong. (2019).

31 Payal Sampat, "Making Clean Energy Clean, Just & Equitable: Earthworks Position Statement," *Earthworks,* blog, April 16, 2019.

32 Sierra Club, "Mining and Mining Law Reform Policy," Sierra Club, February 20, 2020.

33 Michael Braungart and William McDonough, *Cradle to Cradle: Remaking the Way We Make Things* (New York: North Point, 2002).

34 Nicholas Rees and Richard Fuller, *The Toxic Truth: Children's Exposure to Lead Pollution Undermines a Generation of Future Potential* (New York: UNICEF and Pure Earth, 2020).

35 Elsa Olivetti and Jeremy Gregory, Camanoe Associates, *Life Cycle Assessment of Alkaline Battery Recycling,* Report for the Corporation for Battery Responsibility, March 2018.

36 World Bank, "New World Bank Fund to Support Climate-Smart Mining for Energy Transition," press release (Washington, DC: World Bank, May 1, 2019).

37 "Letter to Kristalina Georgieva, CEO, World Bank from a coalition of environmental groups," April 30, 2019, https://earthworks.org/cms/assets/uploads/2019/04/NGOLetterToWorld-Bank_MiningAndRenewables_20190430.pdf.

38 See Supplementary Table 7 of Chengjian Xu et al., "Future Material Demand for Automotive Lithium-Based Batteries," *Communications Materials* 1, no. 1 (December 2020): 1–10.

39 Joannes Mongardini and Aneta Radzikowski, "Global Smartphone Sales May Have Peaked: What Next?" (IMF Working Paper, WP/20/70, May 2020).

40 Shalanda Baker, *Revolutionary Power: An Activist's Guide to the Energy Transition* (Washington, DC: Island, 2021).

41 Channing Arndt et al., "Faster Than You Think: Renewable Energy and Developing Countries," *Annual Review of Resource Economics* 11, no. 1 (October 5, 2019): 149–68.

42 Baker, *Revolutionary Power* (2021).

43 "Just Transition Principles," Climate Justice Alliance, November 2019, https://climatejusticealliance.org/wp-content/uploads/2019/11/CJA_JustTransition_highres.pdf.

BIBLIOGRAPHY

ARCHIVAL AND PRIMARY SOURCE COLLECTIONS

Economic Cooperation Administration. Records of the US Foreign Assistance Agencies, 1948–61. Record Group 469. US National Archives, College Park, MD.

Legislative Commission on Waste Management. Administrative Records, Minnesota State Archives, St. Paul.

Minnesota Pollution Control Agency, Published Records, Minnesota State Archives, St. Paul.

Philipsburg Mining Company Records, 1911–1944. MC 315. Montana Historical Society, Research Center Archives, Helena.

Rayovac Records, 1907–2004. Mss 1170. Wisconsin Historical Society, Madison.

Soo Hardware Company Records, 1911–1918. Collection 2102. Manuscripts and Archives Repository, Hagley Library, Wilmington, DE.

Toxic Docs. Edited by Merlin Chowkwanyun, Gerald Markowitz, and David Rosner. Version 1.0. New York: Columbia University and City University of New York, 2018. www.toxicdocs.org.

United States Geological Survey. *Historical Statistics for Mineral and Material Commodities in the United States*. www.usgs.gov/centers/nmic/historical-statistics-mineral-and-material-commodities-united-states.

United States Geological Survey and US Bureau of Mines. *Minerals Yearbook*, 1885–2020.

The Wayback Machine, Internet Archive. https://archive.org/web/.

INTERVIEWS BY THE AUTHOR

Boolish, Marc. Energizer Brands. Telephone. August 27, 2013.

Coy, Todd. Executive vice president Kinsbursky Brothers International. Anaheim, CA. July 22, 2014.

Huot, Jean-Yves. Telephone. July 31, 2013.

Johnson, Paul. Executive director of environmental affairs at Kinsbursky Brothers International. Anaheim, CA. July 22, 2014.

Marolia, Khush. Global Product Stewardship at Procter & Gamble. Telephone. July 22, 2014.

Scarr, Robert. Telephone. July 22, 2013.

Telzrow, Terry. Telephone. June 21, 2013.

SELECTED PRIMARY AND SECONDARY SOURCES

The following bibliography generally includes primary sources and secondary sources that were cited multiple times. Sources cited only once are found in the notes.

Aronoff, Kate, et al. *A Planet to Win: Why We Need a Green New Deal.* London: Verso, 2019.

Ausubel, Jesse H. "Can Technology Spare the Earth?" *American Scientist* 84, no. 2 (March 1996).

Baker, Shalanda. *Revolutionary Power: An Activist's Guide to the Energy Transition.* Washington, DC: Island Press, 2021.

Barandiarán, Javiera. "Lithium and Development Imaginaries in Chile, Argentina and Bolivia." *World Development* 113 (January 2019): 381–91.

Black, Megan. *The Global Interior: Mineral Frontiers and American Power.* Cambridge, MA: Harvard University Press, 2018.

Braungart, Michael, and William McDonough. *Cradle to Cradle: Remaking the Way We Make Things.* New York: North Point, 2002.

Buchmann, Isidor. *Batteries in a Portable World: A Handbook on Rechargeable Batteries for Non-Engineers.* Richmond, British Columbia: Cadex Electronics Inc., 2011.

Caradonna, Jeremy, et al. "A Call to Look Past an Ecomodernist Manifesto: A Degrowth Critique." 2015. www.resilience.org/stories/2015-05-06/a-degrowth-response-to-an-ecomodernist-manifesto/.

Collantes, Gustavo Oscar. "The California Zero-Emission Vehicle Mandate: A Study of the Policy Process, 1990-2004." Ph.D. diss., University of California, Davis, 2004.

Commoner, Barry. *The Closing Circle: Nature, Man, and Technology.* New York: Random House, 1971.

Crosby, Alfred W. *Children of the Sun: A History of Humanity's Unappeasable Appetite for Energy.* New York: Norton, 2006.

Eckel, William. "The Secondary Lead Smelting Industry." Ph.D. diss., George Mason University, 2001.

Eckel, William P., Michael B. Rabinowitz, and Gregory D. Foster. "Investigation of Unrecognized Former Secondary Lead Smelting Sites: Confirmation by Historical Sources and Elemental Ratios in Soil." *Environmental Pollution* 117, no. 2 (April 2002): 273–79.

Exporting Hazards: U.S. Shipments of Used Lead Batteries to Mexico Take Advantage of Lax Environmental and Worker Health Regulations. San Francisco and Mexico City: Occupational Knowledge International and Fronteras Comunes, June 2011.

Faust, Robert. "Lead Belt Progressives: The Struggle for Social and Environmental Reform in Missouri Mining Communities." Ph.D. diss., University of Missouri–Columbia, 2003.

Fletcher, Seth. *Bottled Lightning: Superbatteries, Electric Cars, and the New Lithium Economy.* New York: Hill and Wang, 2011.

Fredrickson, Leif. "The Age of Lead: Metropolitan Change, Environmental Health, and Inner City Underdevelopment in Baltimore." Ph.D. diss., University of Virginia, 2017.

Giurco, Damien, et al. "Requirements for Minerals and Metals for 100% Renewable Scenarios." In *Achieving the Paris Climate Agreement Goals*, edited by Sven Teske. Cham, Switzerland: Springer, 2019.

Graedel, T. E., et al. "Methodology of Metal Criticality Determination." *Environmental Science & Technology* 46, no. 2 (January 17, 2012): 1063–70.

Greenfield, Aaron, and T. E. Graedel. "The Omnivorous Diet of Modern Technology." *Resources, Conservation and Recycling* 74 (May 2013): 1–7.

Hazleton, Jared E. *The Economics of the Sulphur Industry.* New York: Routledge, 1976.

Hintz, Eric S. "Portable Power: Inventor Samuel Ruben and the Birth of Duracell." *Technology and Culture* 50, no. 1 (January 2009): 24–57.

Huber, Matt T. "Ecological Politics for the Working Class." *Catalyst* 3, no. 1 (Spring 2019).

Hund, Kirsten, et al. *The Mineral Intensity of the Clean Energy Transition.* Washington, DC: World Bank Group, 2020.

Jacobson, Mark Z., and Mark A. Delucchi. "Providing All Global Energy with Wind, Water, and Solar Power, Part II: Reliability, System and Transmission Costs, and Policies." *Energy Policy* 39, no. 3 (March 2011): 1170–90.

Jansson, Johanna. *CSR Practice in the DRC's Mining Sector by Chinese Firms.* Africa Institute of South Africa (AISA), 2010.

Jones, Christopher F. *Routes of Power: Energy and Modern America.* Cambridge, MA: Harvard University Press, 2014.

King, Alexander H. "Our Elemental Footprint." *Nature Materials* 18, no. 5 (May 2019): 408–9.

Kirk, Andrew G. *Counterculture Green: The Whole Earth Catalog and American Environmentalism.* Lawrence: University Press of Kansas, 2007.

Kirsch, David A. *The Electric Vehicle and the Burden of History.* New Brunswick, NJ: Rutgers University Press, 2000.

Klein, Naomi. *This Changes Everything: Capitalism vs. The Climate.* New York: Simon and Schuster, 2014.

Klinger, Julie Michelle. *Rare Earth Frontiers: From Terrestrial Subsoils to Lunar Landscapes.* Ithaca, NY: Cornell University Press, 2018.

Krueger, Jonathan. *International Trade and the Basel Convention.* Washington, DC: Earthscan, 1999.

Latour, Bruno. "An Attempt at a 'Compositionist Manifesto.'" *New Literary History* 41, no. 3 (2010): 471–90.

———. "Love Your Monsters—Why We Must Care for Our Technologies As We Do Our Children." Breakthrough Institute (2011). https://thebreakthrough.org /journal/issue-2/love-your-monsters.

LeCain, Timothy. *Mass Destruction: The Men and Giant Mines That Wired America and Scarred the Planet.* New Brunswick, NJ: Rutgers University Press.

———. *The Matter of History: How Things Create the Past.* New York: Cambridge University Press, 2017.

Levine, Steve. *The Powerhouse: America, China, and the Great Battery War.* New York: Penguin, 2016.

Linden, David, and Thomas B. Reddy. *Handbook of Batteries.* New York: McGraw-Hill, 2002.

Lovins, Amory B. *Soft Energy Paths: Toward a Durable Peace.* New York: Harper and Row, 1977.

MacBride, Samantha. *Recycling Reconsidered: The Present Failure and Future Promise of Environmental Action in the United States.* Cambridge, MA: MIT Press, 2012.

Markowitz, Gerald, and David Rosner. *Deceit and Denial: The Deadly Politics of Industrial Pollution.* Berkeley: University of California Press, 2002.

———. *Lead Wars and the Fate of America's Children.* Berkeley: University of California Press, 2013.

Mascarenhas-Swan, Michelle. "The Case for a Just Transition." In *Energy Democracy: Advancing Equity in Clean Energy Solutions,* edited by Denise Fairchild, 37–56. Washington, DC: Island Press, 2017.

McNeill, John. *Something New under the Sun: An Environmental History of the Twentieth-Century World.* New York: Cambridge University Press, 2000.

Mittlefehldt, Sarah. "From Appropriate Technology to the Clean Energy Economy: Renewable Energy and Environmental Politics since the 1970s." *Journal of Environmental Studies and Sciences* 8, no. 2 (June 2018): 212–19.

Montrie, Chad. *The Myth of Silent Spring: Rethinking the Origins of American Environmentalism.* Berkeley: University of California Press, 2018.

Moyers, Bill D. *Global Dumping Ground: The International Traffic in Hazardous Waste.* Washington, DC: Seven Locks, 1990.

National Research Council. *Effectiveness of the United States Advanced Battery Consortium as a Government-Industry Partnership.* Washington, DC: National Academies Press, 1998.

Nordhaus, Ted, et al. "Ecomodernist Manifesto: A Manifesto for a Good Anthropocene." Last modified 2015. www.ecomodernism.org.

Nye, David E. *Consuming Power: A Social History of American Energies.* Cambridge, MA: MIT Press, 1998.

Olivetti, Elsa, Jeremy Gregory, and Randolph Kirchain. *Life Cycle Impacts of Alkaline Batteries with a Focus on End-of-Life.* Study for National Electrical Manufacturers Association, MIT, February 2011.

Orr, David. *Ecological Literacy: Education and the Transition to a Postmodern World.* Albany: State University of New York Press, 1992.

Pacala, Stephen, and Robert Socolow. "Stabilization Wedges: Solving the Climate Problem for the Next 50 Years with Current Technologies." *Science* 305, no. 5686 (August 13, 2004): 968–72.

Putnam, Hayes & Bartlett. *The Impacts of Lead Industry Economics on Battery Recycling.* Report to the Office of Policy Analysis, US Environmental Protection Agency. Cambridge, MA, 1986.

———. *The Impacts of Lead Industry Economics and Hazardous Waste Regulation on Lead-Acid Battery Recycling: Revision and Update.* Report to the Office of Policy Analysis, US Environmental Protection Agency. Cambridge, MA, September 1987.

Rohr, Johannes. *Indigenous Peoples in the Russian Federation.* Copenhagen, Denmark: International Work Group for Indigenous Affairs, 2014.

Rome, Adam. "Crude Reality." *Modern American History* 1, no. 1 (January 5, 2018): 1–6.

Ruble, Kenneth D. *The RAYOVAC Story: The First 75 Years.* Madison, WI: Rayovac, 1981.

Schallenberg, R. H., *Bottled Energy: Electrical Engineering and the Evolution of Chemical Energy Storage.* Philadelphia: American Philosophical Society, 1982.

Scharff, Virginia. *Taking the Wheel: Women and the Coming of the Motor Age.* Albuquerque: University of New Mexico Press, 1992.

Schlesinger, H. R. *The Battery: How Portable Power Sparked a Technological Revolution.* Washington, DC: Smithsonian, 2010.

Schumacher, E. F. *Small Is Beautiful: Economics as If People Mattered.* New York: Harper Perennial, 1973.

Sellers, Christopher C. "Cross-Nationalizing the History of Industrial Hazard." *Medical History* 54, no. 3 (2010), 315–40.

———. *Hazards of the Job: From Industrial Disease to Environmental Health Science.* Chapel Hill: University of North Carolina Press, 1999.

Shaner, Matthew R., et al. "Geophysical Constraints on the Reliability of Solar and Wind Power in the United States." *Energy & Environmental Science* 11, no. 4 (April 18, 2018): 914–25.

SmithBucklin Corporation. *National Recycling Rate Study.* Reports prepared every few years for Battery Council International. Chicago, 1995–2017.

Steinberg, Theodore. *Down to Earth: Nature's Role in American History.* New York: Oxford University Press, 2002.

Turner, James Morton. *The Promise of Wilderness: American Environmental Politics since 1964.* Seattle: University of Washington Press, 2012.

Turner, James Morton, and Andrew C. Isenberg. *The Republican Reversal: Conservatives and the Environment from Nixon to Trump.* Cambridge, MA: Harvard University Press, 2018.

Urry, John. *Mobilities.* Malden, MA: Polity, 2007.

US Department of Energy, "Critical Materials Strategy." Washington, DC: US Department of Energy, December 2011.

US Geological Survey. "Historical Lead Statistics." Last modified December 20, 2018. www.usgs.gov/centers/nmic/historical-statistics-mineral-and-material-com modities-united-states.

Vidal, Olivier, Bruno Goffé, and Nicholas Arndt. "Metals for a Low-Carbon Society." *Nature Geoscience* 6 (November 2013): 894–95.

Vinal, George. *Storage Batteries: A General Treatise on the Physics and Chemistry of Secondary Batteries and Their Engineering Applications.* 3rd ed. New York: Wiley and Sons, 1940.

Warren, Christian. *Brush with Death: A Social History of Lead Poisoning.* Baltimore: Johns Hopkins University Press, 2000.

Worster, Donald. *The Wealth of Nature: Environmental History and the Ecological Imagination.* New York: Oxford University Press, 1993.

Xu, Chengjian, et al. "Future Material Demand for Automotive Lithium-Based Batteries." *Communications Materials* 1 (December 2020): 1–10.

Zimring, Carl A. *Cash for Your Trash: Scrap Recycling in America.* New Brunswick, NJ: Rutgers University Press, 2005.

INDEX

US Consumer Product Safety Commission, 127

US Signal Corps, 72

vehicle, conventional: dry-cell starter batteries for, 61; fuel efficiency standards for, 140, 155–56; invention of electric starter for, 19; lead-acid starter battery for, 13, 20; policy goals to phase out, 2, 158, 166. *See also* electric vehicles; hybrid vehicles

vehicle, electric. *See* electric vehicles

vehicle, hybrid. *See* hybrid vehicles

Vermont, 58, 89

Vernon, California, 47–48

Vidal, Olivier, 7

Volkswagen (VW), 2, 145, 157

Volt (Chevrolet), 137, 140, 155, 156, 161

Walkman, Sony, 9, 80, 101

Wall Street Journal, 140

Wang, Chuanfu, 124–25

Warden, Jack, 46

Warden, Leslie, 46

Warren, Christian, 26–27

Washington, Booker T., 22

Whittingham, Stanley, 98–99, 101, 129

Who Killed the Electric Car? (documentary), 16, 135

wilderness, 170

Willard Storage Battery Company, 21, 24–27, 33

wireless resolution. *See* smartphone

World Bank: Climate-Smart Mining initiative, 182–83; DRC mining code, 115

World Health Organization, 53

World War I, 68

World War II, 65, 66, 71–73

XPeng (Chinese automaker), 157

Yoshino, Akira, 97, 100–102, 119, 129

Young, Alison, 46

zero-emissions vehicles. *See* electric vehicles

zinc, 77, 79, 83, 161–62

zinc-carbon batteries, 59, 60–61, 73; assembly of, 92–93; compared to alkaline-manganese, 78–80, 81; disposal of, 82–83; material history of, 65–70, 76–77; performance of, 65, 71, 77. *See also* batteries, disposable

WEYERHAEUSER ENVIRONMENTAL BOOKS

Charged: A History of Batteries and Lessons for a Clean Energy Future, by James Morton Turner

Wetlands in a Dry Land: More-Than-Human Histories of Australia's Murray-Darling Basin, by Emily O'Gorman

Seeds of Control: Japan's Empire of Forestry in Colonial Korea, by David Fedman

Fir and Empire: The Transformation of Forests in Early Modern China, by Ian M. Miller

Communist Pigs: An Animal History of East Germany's Rise and Fall, by Thomas Fleischman

Footprints of War: Militarized Landscapes in Vietnam, by David Biggs

Cultivating Nature: The Conservation of a Valencian Working Landscape, by Sarah R. Hamilton

Bringing Whales Ashore: Oceans and the Environment of Early Modern Japan, by Jakobina K. Arch

The Organic Profit: Rodale and the Making of Marketplace Environmentalism, by Andrew N. Case

Seismic City: An Environmental History of San Francisco's 1906 Earthquake, by Joanna L. Dyl

Smell Detectives: An Olfactory History of Nineteenth-Century Urban America, by Melanie A. Kiechle

Defending Giants: The Redwood Wars and the Transformation of American Environmental Politics, by Darren Frederick Speece

The City Is More Than Human: An Animal History of Seattle, by Frederick L. Brown

Wilderburbs: Communities on Nature's Edge, by Lincoln Bramwell

How to Read the American West: A Field Guide, by William Wyckoff

Behind the Curve: Science and the Politics of Global Warming, by Joshua P. Howe

Whales and Nations: Environmental Diplomacy on the High Seas, by Kurkpatrick Dorsey

Loving Nature, Fearing the State: Environmentalism and Antigovernment Politics before Reagan, by Brian Allen Drake

Pests in the City: Flies, Bedbugs, Cockroaches, and Rats, by Dawn Day Biehler

Tangled Roots: The Appalachian Trail and American Environmental Politics, by Sarah Mittlefehldt

Vacationland: Tourism and Environment in the Colorado High Country, by William Philpott

Car Country: An Environmental History, by Christopher W. Wells

Nature Next Door: Cities and Trees in the American Northeast, by Ellen Stroud

Milton Keynes UK
Ingram Content Group UK Ltd.
UKHW010631110724
445408UK00004B/151